Is There a Duty to Die?

Edited by

James M. Humber and Robert F. Almeder

BIOMEDICAL ETHICS REVIEWS

Is There a Duty to Die?

Edited by

James M. Humber

and

Robert F. Almeder

Georgia State University, Atlanta, Georgia

Humana Press • Totowa, New Jersey

999 Riverview Drive, Suite 208
Totowa, NJ 07512

For additional copies, pricing for bulk purchases, and/or information about other Humana titles, contact Humana at the above address or at any of the following numbers: Tel.: 973-256-1699; Fax: 973-256-8341; E-mail: humana@humanapr.com, or visit our Website: http://humanapress.com

This publication is printed on acid-free paper. ∞

ANSI Z39.48-1984 (American National Standards Institute) Permanence of Paper for Printed Library Materials.

Cover design by Patricia F. Cleary.

ISBN 0-89603-783-5

Printed in the United States of America. 10 9 8 7 6 5 4 3 2 1

The Library of Congress has cataloged this serial title as follows:

Biomedical ethics reviews—1983– Totowa, NJ: Humana Press, c1982–
v.; 25 cm—(Contemporary issues in biomedicine, ethics, and society)
Annual.
Editors: James M. Humber and Robert F. Almeder.
ISSN 0742-1796 = Biomedical ethics reviews.
1. Medical ethics—Periodicals. I. Humber, James M. II. Almeder, Robert F.
III. Series.
[DNLM: Ethics, Medical—periodicals. W1 B615 (P)]
R724.B493 174'.2'05—dc19 84-640015
AACR2 MARC-S

Contents

Preface

The question of whether there might be a duty to die was first raised by Margaret Battin in 1987 in her ground-breaking essay, "Age Distribution and the Just Distribution of Health Care: Is There a Duty-to-Die?" In 1997 the issue was reprised when two new articles appeared on the topic written by John Hardwig and the other by former Colorado Governor Richard D. Lamm. Given the renewed interest in the topic, as well as its undeniable importance, *Biomedical Ethics Reviews* sought to initiate an in-depth discussion of the issue by soliciting articles and issuing a general call for papers on the topic "Is There a Duty to Die?" The twelve articles in this volume represent the ultimate fruits of those initiatives.

The first seven essays in this text are sympathetic to the claim that there is a duty to die. They argue either: (a) that some form of a duty to die exists, or (b) that arguments that might be offered against the existence of such a duty cannot be sustained. By way of contrast, the last five articles in the text are critical of duty-to-die claims: The authors of the first three of these five articles attempt to cast doubt on the existence of a duty to die, and the writers of the last two essays argue that if such a duty did exist, severe problems would arise whenever we attempted to implement it.

"Is There a Duty to Die?" is the seventeenth annual volume of *Biomedical Ethics Reviews*, a series of texts designed to review and update the literature on issues of central importance in bioethics today. For the convenience of our readers, each article in every issue of *Biomedical Ethics Reviews* is prefaced by a short abstract describing that article's content. Each volume in the series is organized around a central theme; the theme for the next volume of *Biomedical Ethics Reviews* will be "Privacy and Health Care." We hope our readers will find the present volume of *Biomedical Ethics Reviews* to be both enjoyable and informative, and that they will look forward with anticipation to the publication of "Privacy and Health Care."

James M. Humber
Robert F. Almeder

Contributors

Michael Almeida • Department of English, Classics, Philosophy, and Communication, The University of Texas at San Antonio, Texas

Susan Leigh Anderson • Department of Philosophy, University of Connecticut at Stamford, Connecticut

Margaret P. Battin • Department of Philosophy, University of Utah, Salt Lake City, Utah

Marilyn Bennett • Department of Philosophy, College of St. Catherine, St. Paul, Minnesota

J. Angelo Corlett • Department of Philosophy, San Diego State University, San Diego, California

David Drebushenko • Department of Philosophy and Political Science, University of Southern Indiana, Evansville, Indiana

Robert E. Ehman • Department of Philosophy, Vanderbilt University, Nashville, Tennessee

Judith Lee Kissell • Interdisciplinary Studies, Georgia College and State University, Milledgeville, Georgia

Paul T. Menzell • Office of the Provost and Dean of Graduate Studies, Pacific Lutheran University, Tacoma, Washington

Jan Narveson • Department of Philosophy, University of Waterloo, Ontario, Canada

Ryan Spellecy • Department of Philosophy, University of Utah, Salt Lake City, Utah

Rosemarie Tong • Department of Philosophy, The University of North Carolina at Charlotte, North Carolina

Abstract

Is there a duty to die? Consider the stark differences in life expectancy around the world, from as high as 80 in the richest nations to below 50 in the poorest. There are also stark global differences in access to health care. However, despite these stark differences, many theorists of distributive justice will reject the claim that inhabitants of the rich countries owe those of the poor countries anything at all, and certainly have no duty to die to conserve and redistribute health care resources in order to even these life expectancies out.

I explore arguments by John Hardwig about duties to one's family, and, against a Rawlsian background explored by Norman Daniels, arguments by Margaret Battin and Dan Callahan concerning duties in a larger social setting. I suggest that although these arguments initially seem not to support any such "duty to die" for reasons of global equity, they can plausibly be stretched to do so. In the end, what blocks the current existence of any global "duty to die" is the lack of global redistributive structures that would convey the savings from one person's earlier death in the first world—whether the result of declining expensive life-prolonging treatment or physician-assisted suicide—to fund health care and related measures that would increase life expectancies in the second and third worlds, as well as a failure to recognize the existence of mutual health-related obligations between rich nations and poor ones. I argue that there may well be a moral obligation to develop these redistributive structures against a background of mutual obligation, which then would underwrite a more general "duty to die" for reasons of global equity.

Is this argument a reductio ad absurdum*, or is it a "highly demanding" one? This paper closes with attention to this question, and makes some predictions about the evolution of moral obligations in the future.*

Global Life Expectancies and the Duty to Die

Margaret P. Battin

Is there a duty to die? This inflammatory question, often originally attributed to then-Governor of Colorado Richard Lamm, has been explored within the context of American health care both as a personal issue and as a societal one. Recently, John Hardwig has posed it in the highly personal context of a troubling rumination about duties to his own family—duties he says he willingly accepts not to burden them with obligations of excessive expense or care as he succumbs to extreme old age or terminal illness.[1] Some years earlier, drawing in part on work by Norman Daniels, I had posed the same question in an impersonal, social context about the choices one might rationally make under conditions of moderate scarcity in access to health care.[2] However, neither Hardwig nor I have explored the question in a still larger context, that context in which it may seem to be both least persuasive to some, but most troubling to others.[3] This is the context of global justice.

Is there a duty to die? Consider the stark differences in life expectancy around the world. In the rich, industrially developed nations, average life expectancy ranges roughly between 72 and 80 for both sexes, with Japan, Canada, Iceland, and the Netherlands at the top end of the range; in the poorest, not-yet-developed agrarian nations of the so-called third world, in contrast, life expectancy ranges downward from 60 to 50, and in some cases, like Equatorial Guinea, Zambia, Angola, Uganda, and Sierra Leone, among other sub-Saharan African countries, below 50.[4] Although life expectancy has been increasing in most nations, in some, like Russia in the post-Soviet years, it is plummeting, down from 69.2 in the 5-yr period 1985–1990 to an estimated 64.4 for 1995–2000, with men dropping to 58.0.

There are also stark global differences in access to health care. The high-income countries (those with per capita annual incomes above $8500) get almost all the health care made available in the world: in 1994, for example, the rich countries accounted for 89% of global health expenditures, even though they comprise only 16% of the global population.[5] Furthermore, of the estimated 1.4 trillion disability-adjusted years of life that were lost to disease in that year, the inhabitants of the rich nations suffered just 7% of them.[6] People in rich countries live far longer, far healthier lives, and die much, much later than people in poor countries.

Can these differences serve to raise a question about a duty to die? Perhaps, like most other readers of these remarks, you will reject the question out of hand, even if moved by the plight of distant peoples around the globe. Although life expectancies may be unequal, you will no doubt say that this hardly establishes that they are inequitable, or that those with longer life expectancies have any duty to those with short life expectancies to even things out. To recognize that life expectancies, like infant mortality and incidence rates for various diseases, are comparatively crude measures of morbidity and mortality, and that there are other far more sophisticated measures of health status and well-being may still not persuade you that there is any inequity in health between the rich and poor nations that we ought to remedy; we do not, you may say, owe them anything at all.[7] No doubt you will agree that it is unfortunate that inhabitants of the poor countries receive less health care, experience much worse health, and lose far more years of life, but even if you grant that it is unfortunate, you will be less likely to see it as unfair, and in any case, you almost certainly do not think that global differences in patterns of life expectancy can impose a duty to *die* on the residents of more fortunate nations.[8]

However, if we look at the various arguments that both Hardwig and I (as well as others on whom we have drawn or who hold related views, like Norman Daniels and Dan Callahan) have made about whether there might be a duty to die in what we all assumed was a western, first-world, American context, we may see that these arguments have implications reaching far beyond our borders—implications yielding troubling results in a global context—and perhaps bad news for you and me.

Arguments About the Duty to Die

Hardwig's Argument

Hardwig argues that an individual who is terminally ill or in need of extensive care may have a duty not only to decline this care, but to die, in order to avoid imposing overly heavy burdens of care and support on family members or loved ones. This duty Hardwig sees as being stronger for people who are older and who have already lived full lives, especially if they are facing dementing disease like Alzheimer's or Huntington's, whose loved ones have had difficult lives or have already made sacrifices for them, and who can no longer hope to make significant contributions to the lives of their loved ones. An individual in this situation—and Hardwig clearly accepts the prospect that he may someday find himself in this position—ought to be willing to die in order to avoid "stealing the futures" of his or her loved ones, who would otherwise shoulder duties of care to him or her, rather than buying a little more time for himself or herself.[9]

This is a complex and controversial argument, but given that it addresses obligations only to immediate family members and loved ones who are directly and severely affected by one's remaining alive, Hardwig's argument would seem to have no implications concerning distant peoples, remote residents of the world whom one does not know, with whom one does not interact, and who inhabit utterly different cultural, social, and religious spheres from one's immediate family. There are no personal ties here, no relationships of mutual affection, no obligations assumed in marriage or in traditional filial relationships, not even any relationships of co-nationality, and although we think it might be a good thing, a decent thing to contribute some of one's assets to help these unfortunate people, there is certainly no duty to sacrifice one's life.

Battin's Argument

My own argument about duties to die has not been driven by sentiments of affection or concern for family members, nor is it confined to the orbit of the intimate family situation. Rather, one might say, this argument has been driven by cold self-interest, at least of a theoretical sort. I had claimed that rational self-interest maximizers in the original position (whom one might recognize from Rawls[10]), choosing principles that would govern a society characterized by moderate scarcity of health care resources, would (as Norman Daniels had argued[11]) recognize that health care resources spent earlier in life—that is, on people in younger age groups—would both be more efficient and would raise the chances of survival in early life (a precondition for later life), thus (except in the first generation of such a policy) enhancing one's chances of longer, better survival at a later stage of life. Thus, parties in the original position would allocate health care away from the oldest ranges toward the youngest ranges of everyone's life-spans, and hence, age-rationing policies would count as just. However, if there were a duty to refrain from using resources in terminal illness or extreme old age that might be more justly be allocated to the care of younger people, I then argued, given the potentially cruel consequences of living without adequate care, many rational self-interest maximizers in the Rawlsian original position would welcome social policies recognizing elective assisted suicide or active voluntary euthanasia—not obligatory or forced, not age-tagged, and certainly not secret, but available for those who might choose them. Thus, there would not be a duty to die in itself, but if there were an obligation to refrain from receiving care, dying might seem the preferable choice. Thus, policies permitting or even encouraging voluntary choices of an earlier death, if supported in the original position, would then be just.

This, too, is a complex argument, and it, too, may seem to have little or no relevance to the global context. After all, it is not clear how voluntary choices of an earlier death, including those involving physician-assisted suicide or physician-performed euthanasia, even if supported by social practices and expectations, would have real impact on global differences in life expectancy. Drawing on data from the Netherlands, we can expect that such choices, at least as made in a current world in which they are legally tolerated and fairly widely socially accepted, would mean forgoing only the last few weeks of life; only comparatively few people would make them, and the savings in health care costs would be modest, less than 0.07 percent of annual total health care expenditures in the United States.[12] Such choices would have comparatively little impact on overall health care costs even in a developed nation that permitted them, and virtually no discernible impact on global differences in life expectancy of people in poor countries.

Callahan's Argument

Meanwhile, Dan Callahan has also been making use of arguments that, in their origins, have been sympathetic to Daniels' view of age-rationing. Callahan's initial view was that the elderly should rethink the meaning of old age and refrain from claiming expensive health care resources in an attempt to prolong life indefinitely; he has expanded this view to insist that all of us reinspect our assumptions about health, disease, aging, and death, and abandon our relentless pursuit of medical "progress" in general, not only in old age.[13] We must stop assuming that we can conquer all disease, or indefinitely prolong life.

Does Callahan think there is a duty to die? He is staunchly opposed to any direct ending of life, and would certainly deny that there is any "duty to die" in the sense accepted straightforwardly by Hardwig or partially and obliquely by myself. However, he recognizes that the rethinking of cultural assumptions about medical progress and the meaning of life will often mean that death arrives earlier, and he clearly recognizes obligations in connection with the approach of death: this is a "passive" duty, the duty of restraint, an obligation to refrain from claiming expensive health care resources that might postpone death and prolong one's life.

Of these three discussions of duties at the end of life, Callahan's alone is sensitive to global issues in health. He holds that greater equality in health circumstances ought to be brought about around the world, but he sees this simply as a matter of holding back in the first world from our unthinking dedication to so-called medical progress, holding back in a way that will allow us to achieve a "steady-state" medicine, allowing the developing world to catch up gradually with the developed world, so that global health care equality will be more nearly achieved. He does not explore the more direct obligations that his view, like that of Hardwig, Daniels, and Battin, might seem to support.

The Global Analogy

These views about duties to die or to refrain from care may seem to have little to do with the global situation or with differences in life expectancies around the world, differences that remain (though not as starkly) even if high infant mortality rates are factored out. That some sub-Saharan Africans have life expectancies under 50 or Russian men have life expectancies that have dropped to 58, whereas those of Japanese women have climbed to 83.3 (more than double the 41.0 of people in Malawi) cannot, it may seem, generate obligations on the part of inhabitants of richer nations—especially not substantial, personal obligations like a duty to die. After all, the Hardwig, Battin, and Callahan views about duties to die, whether accepted, accepted obliquely, or rejected, are all predicated on Daniels' assumption that savings gained in the costs of care if someone dies sooner have effects in other immediate contexts—for Hardwig, because they spare costs affecting the nonfinancial and financial futures of family members, for Battin and Callahan because they can be utilized in care for younger age groups.

However, as Daniels has pointed out, such arguments trade on the assumption that savings are made in a closed system, one in which savings in one part yield usable resources in another.[14] The Rawlsian argument as adapted by Daniels to the health care setting, on which the Battin and Callahan arguments in particular trade, presupposes not only a background of just institutions, but redistributive mechanisms that would ensure that savings in one place, namely health care in old age, would be reallocated somewhere else, namely as health care for the young. Closed health care financing systems like this are imperfectly in place in some nations, like Canada (and like Britain before the National Health System began to be partly dismantled), but not in some other developed nations, particularly the United States.

However, the developed, post-Soviet, and the developing nations—the "first," "second," and "third" worlds—do not jointly form a closed system, in which health care savings in one part of the world are realized as resources in another. To decline dialysis in England does not buy immunization against malaria in Brazil; to have one's physician assist in suicide before the end stages of cancer in the United States does not curb tuberculosis in Russia or reduce perinatal mortality in India. To be sure, declining dialysis or committing suicide would have some impact on reducing consumption rates in the developed world, inasmuch as someone whose death occurs earlier no longer consumes food, energy, or natural resources, but the extent of this impact would be negligible in altering medical outlooks and life expectancies in the developing world. Nor are we participants in a global cooperative scheme, one that might generate obligations of mutual support for each other among the various inhabitants of the globe: there do not seem to be special obligations generated toward each other within a cooperative system.[15] Thus, duty to die arguments, even if plausible within a single nation-state with an interactive health-care system, seem to have no purchase in the global context, where, on the contrary, the connection between a patient in the developed world and distant peoples in the developing world is so remote that the effects of an earlier death in one place will be barely detectable if at all in the other—so infinitesimally small that it could hardly seem worth the sacrifice of a life.

Of course, I might draw up a will conveying my estate to the residents of, say, a specific African village, but unless I do so, my death—whether earlier or later—will have no impact on the fates of these distant peoples. My own earlier death would not protect Russian men nor for that matter would it do anything to slow the climb of Japanese women up the scale of long life expectancy. This is not to say that earlier, cheaper deaths cannot benefit others. John Hardwig's earlier death, if he can will himself to bring it about, might well protect his family members from burdens of expense and care they would shoulder out of duty to him. In such a manner might my own earlier death benefit my own similarly immediate family and, depending on the particular sort of insurance system by which I am covered, other claimants for health care within the same system as well. Callahan's refusal to claim life-prolonging resources or to succumb to the dream of indefinite medical progress, if he can persuade others as well as himself to do as he recommends, would benefit American society as a whole. All three of our earlier and thus less expensive, more resource-conserving deaths might save health care resources for redistribution within the American health care system, at least within the immediate insurance pools of which we are each part, to provide for health care for some younger persons and thus give them a greater chance of longer life. But nothing Hardwig or Callahan or I might do in shortening our lives or letting them be shorter would seem to affect the welfare of people in distant countries, or in any way serve to decrease the disparity between their lifespans and ours. Arguing that John, Dan, I, or, for that matter, you, the reader of this paper, ought to die a little earlier to extend the life-spans of people in the poorer nations of Africa, Asia, or Latin America may seem just as futile as insisting that we eat all the food on our plates to keep people from starving in Somalia or the Sudan, or whatever is our current Armenia.

There are of course many things John, Dan, you, and I could do to improve the life prospects of peoples with lower life expectancies than our own. These include whatever contributions we might make to remedying the social and economic circumstances that contribute to early death: reducing malnutrition, endemic disease, poverty, and civil warfare in Africa; controlling the sex trade in southeast Asia; combatting alcoholism, environmental toxins, and exposure to freezing cold in Russia, as would reducing our own consumption in a significant way, and even more directly, sending food, medical supplies, and other aid. All these things would certainly help, and one can certainly claim that it would be a good thing, indeed perhaps a duty, for us to do them.[16] It would produce a fairer world, one in which the life prospects of all global inhabitants were more nearly equal and in which situations beyond those individuals' control did not cause them such harm.

However, these things are not accomplished by any of us seeing ourselves as having a duty to die. Is there a "global" duty to die, or a duty to die for reasons of global health equity? Is seems that the answer is no.

How to Stretch These Arguments to Cover the Globe

There are several assumptions here that can be challenged, however. Neither Hardwig's, Battin's, nor Callahan's arguments about a duty to die, whether positive or negative, seem to have any purchase in a global context, but exposing these assumptions may suggest that they could on the contrary readily be stretched to global scope. I want to look at two of these.

First, in a view most evident in Hardwig's account, we tend to make the non-Kantian, non-Utilitarian assumption that it is close, personal relationships that count, and that generate duties to each other. Certainly the impact of one's continued living may affect one's immediate family more strongly by virtue of their greater proximity, but this need not necessarily be the case. Suppose Hardwig, like the example he describes, were at age 87 to develop congestive heart failure, and his APACHE score predicted that he has less than a 50% chance to live for another six months.[17] He might beg for the most aggressive life-prolonging treatment possible, and indeed, it might succeed in continuing his life for another two years, but this care costs, let us suppose, $100,000 a year and will impose immense burdens of care on his 55-year-old daughter. John is, in his own view, obliged to spare her this, but now imagine the $200,000 for John's health care for the two years he, like his example, manages to survive translated to, say, a village in Uganda, where it will purify the water supply, provide vaccines for all the village children, hire a health care worker to treat wounds, provide emergency surgical and psychiatric care, rehydrate children with diarrhea, and provide family planning and reproductive health care. John's $200,000 would go a very long way here. Indeed, let us suppose, these funds might raise the life expectancy of these villagers by, say, 5 years on average. There are a hundred people in the village, let us say; the total life-years gained would be 500. John, on the other hand, sacrifices just 2. Should he do it? If he ought to die a little earlier to spare his 55-year-old daughter her savings, her home, her job, and her career, should he not do so to spare a hundred people from death 5 years earlier on average than it might otherwise occur? Exactly what difference does it make that one is his own daughter, and the others are people not personally known to him?

It might be replied that Hardwig's argument has involved only sparing burdens, not providing benefits, and that duties of nonmaleficence are far stronger than those of beneficence. However, it could also be argued that given the underlying picture of balancing benefits and burdens that informs Hardwig's conception of family discussions concerning whether one of its members ought indeed to die, there is a more Utilitarian metric at work here than Hardwig acknowledges, one that would underwrite the provision of benefits as well as burdens.[18] If by dying a little earlier the terminally ill family member could, for example, release many thousands of dollars of life insurance just in time for a granddaughter who could not otherwise go to college, or the surviving spouse to rescue a business that would otherwise fail, ought he do it?[19] Hardwig does not address the question of providing benefits by dying earlier specifically, but because he would require the dying person not to impose burdens that would prevent the same sorts of family disadvantages, it may not be out of keeping with his view.

Second, arguments like Battin's and to some extent Callahan's trade on the traditional Rawlsian assumption that limits the range of possibilities one might occupy in the original position to one's own (liberal) society. Parties to the original position recognize, under the veil of ignorance, that they might occupy any position in the society; this is why they do not favor social policies that disadvantage some (e.g., slaves) but benefit others (e.g., slaveholders); after all, they might turn out to be slaves rather than masters. However, the original Rawlsian conjecture was conceived of as confined to a single society, and not thought of in terms of the huge range of societies inhabiting the current world; this was primarily because the conjecture requires the possibility of a moral community in which the benefits and burdens of social cooperation are shared. Now, however, there is an extended literature by Rawls and others on whether this conjecture must remain limited to a nation-state or could be extended to global scope as "the law of peoples."

Though Rawls does not apply this extended conjecture directly to individuals, it is nevertheless plausible to ask why the participants in an original position addressed primarily to matters of health, as in Daniels' conjecture, could not reflect on whether they might be, say, Japanese, with a life expectancy of 79.9, or Burundian, with a life expectancy of 44.5. If they do so, they will surely elect principles of health care distribution and other health-related matters that more greatly favor Burundians than Japanese. Re-examining both arguments in a broader context, it appears that they may well show that the rational self-interest maximizer, recognizing that he or she does not know whether he or she will be Burundian or Japanese, would even favor policies that have the consequence that Japanese live for a slightly shorter time if this permits Burundians to live longer. This will also give the greatest benefit to the least well off. Therefore, with adequate protections against coercion and other sorts of abuse, they might even favor policies that would not only limit our access in the first world to expensive life-prolonging care, but also permit or even encourage us to seek voluntary physician-assisted suicide or euthanasia. "We ought to die sooner, so that they could live longer" may seem unjust, but if we did not know who the "we" and the "they" are and which group we find ourselves among, we will find more nearly equal life-spans far more just.

However, you will surely say this is preposterous. After all, not only would my sacrifice of my life be a clearly futile gesture, since it would have so little impact, but even Daniels' strategy of considering issues of justice over a life-span does not work in the global case: abbreviating my own life would not provide the theoretical advantage, central in Daniels' argument concerning what policies are just, of having already increased my own chances of survival through all my earlier years, all prerequisite to my current existence. This cannot be the case in the kind of world we have, a world in which various nations' health care systems are not interrelated and do not form anything approaching a closed system in which changes in my prospects may enhance yours.

However, seeing this uncovers our central mistake. Our failure to find plausible an argument for a duty to die based on current differences in global life expectancies does not rest in an adequate philosophical objection and does not undermine the assumptions of any of these arguments. Rather, it results from a failure of vision, of seeing accurately what is the case in the world. It is what is actually the case in the world—or fails to be the case in the world—that controls our response here. For what is actually the case in the world is a lack, the lack of actual redistributive mechanisms that might convey the savings from one person's earlier death in the first world to fund health care and related measures that would increase life expectancies in the second and third worlds. Because this is a lack, an absence, it is harder to notice than if there were some more concrete obstacle to global health care justice at hand.

Because of this lack, our arguments about a global duty to die do not succeed now in a world that does not have such structures. But it *could* have them—say, international health care structures that worked to redistribute health care resources from rich to poor nations and thus to more nearly equalize health care resources for all, in an effort to make the life prospects of all humans more nearly equal. Of course, libertarians and antiegalitarians will shrink at the suggestion of such redistributive structures, but egalitarians and Rawlsians will welcome them. After all, Rawls himself argues that regarding societies now "burdened by unfavorable conditions"—he has in mind "the poorer and less technologically advanced societies" of the third world, often burdened by "oppressive government and corrupt elites, the subjection of women abetted by unreasonable religion, with overpopulation relative to what the economy of the society can decently sustain"[20]—the goal is that eventually each of these societies "should be raised to, or assisted toward, conditions that make a well-ordered society possible."[21] The well-ordered and wealthier societies have duties and obligations to societies burdened by such conditions to aid in seeing that human rights are to be recognized and basic human needs are met; this is a "duty of assistance,"[22] and (to extend Rawls' concern to the health-related issues that are our specific focus here) such duties of assistance—moral duties, but duties that might necessitate legal requirements if they are to have any effective force at all—[23]would certainly include the provision of information, technology, and medical resources directly bearing on the health status and life prospects of the inhabitants of these poorer parts of the world.

For consider some of the inequities that currently affect life prospects of the first, second, and third worlds. In the developed countries, access to vaccines is nearly universal: children routinely receive immunizations against tetanus, measles, and polio; in the third world, these are generally unavailable, and death rates from these conditions are high. Or consider reproductive health care: for a large catalog of reasons, maternal mortality is just 7 per 100,000 live births in Sweden, but 1400 per 100,000 in Yemen, 1500 in Nepal, 1600 in Guinea, and 1800 per 100,000 in Sierra Leone.[24] Rabies exposure is treated with comparatively convenient, painless drugs in developed countries; in much of the third world, it has been treated with earlier, less effective drugs, involving much more frequent and painful injections with a much poorer outcome. Or yet again consider people with HIV or AIDS: Patients in the developed countries are now seeing their lives extended by protease inhibitors; those in the third world have none. These inequalities all affect life expectancy in substantial ways. Of course, these inequalities could not now be reduced if John's death occurred a little earlier and more cheaply, or if mine did, too, since what neither of us could guarantee is that the savings we yielded would be translated into longer life for someone else who would otherwise have had dramatically less.

The savings from our earlier deaths would not be large in the first world, but they would have immense impact if they were translated into resources in the third world. If global health care were indeed an interactive, interrelated global system and if the appropriate redistributive structures were in place, these tiny savings from first world choices of earlier death would prove far more efficient in protecting and extending life in the third world: The costs of one expensive and not very effective unit of life-prolonging care in the first world buys many, many units of inexpensive and highly effective care in the second and especially third (vaccination, oral rehydration, basic reproductive health care). This is just the point about savings and efficiency that is so central to the redistributive claims here.

This argument depends on the existence of redistributive structures that function on a global scale, making it possible for savings in the first world to serve as resources in the third. We do not have these structures yet—at least, not in full. But we do have rudimentary and partial systems—rudimentary and partial in that they cover only some conditions or some areas, or are financed in limited ways: such organizations as *Medécins Sans Frontières* and a wide range of partly volunteer, charity-financed organizations that bring first-world medical resources to the third world. Several major foundations, particularly the Rockefeller Foundation, have supported international health initiatives generously. There are even quasi-governmental organizations, like the World Health Organization. These are all financed primarily by the first world, but they are not yet financed by health care savings in the first world. The existence of such institutions would be crucial to the argument at hand.

The argument, at least in its Rawlsian framework, also depends on the assumption that there is a network of interrelated obligations of mutual support that form the basis of a community, a global cooperative scheme. Such obligations cannot be just one way; it cannot be the case that the rich nations have obligations to the poor nations, but not the other way around. To suggest that people in the rich, first-world nations have health-related obligations to people in the second and third worlds that might be reflected in more nearly equalizing life expectancies is thus to raise the issue of whether people in the second and third worlds have health-related obligations to those in the first world. However, we do assume this, we in the rich nations, though these obligations are usually discussed in the context of environmentalism: we assume that the people of the poor nations ought not do things that potentially affect our health and perhaps indeed survival, especially those that damage the environment. Thus, we insist that they ought not cut down the rainforest, they ought not use coal-fired manufacturing processes without adequate pollution controls, they ought not dump pesticides into the sea, they ought not let epidemics of infectious disease run rampant without controls, and they ought not destroy environments that harbor rare species of plants and animals that might provide discoveries of enormous medical and pharmaceutical importance. Of course, these obligations may not be very well recognized in practice—but at least partly because observing them in practice would be at least as costly in terms of their own life prospects for people who depend on clearing new tracts of land, on cheap manufacturing techniques, and so on, as sacrificing an amount of life at the end of a terminal illness would be for us. Nevertheless, we do recognize, however dimly, a network at least in theory of interrelated obligations of mutual health-related support, both not to damage each others' health or potential health and, as argued here, to provide positive measures for improving it.

With these two further components in place, the development of international redistributive structures and a background of interrelated obligations of mutual health-related support in a global cooperative scheme, the conditions would be satisfied within which a "duty to die" could become a reality—that is, a duty to conserve health care resources by forgoing treatment or directly ending one's life in the interests of justice in health care, reflected in more nearly equal health prospects and life expectancies around the globe. Those who argue as Hardwig, Daniels, Battin, and Callahan do, in their various ways, may also be committed to arguing as well that having one's death occur earlier, whether directly caused or as the result of refraining from claiming expensive life-prolonging care, would be the morally right thing—indeed, the just thing to do. I cannot argue fully for that conclusion here, but I do think we should all be on notice that the stakes may be this high—both for those in other parts of the world whose life expectancies are much shorter than ours, and for us whose life expectancies are so comparatively long.

Conclusion

I have tried to show that the arguments Hardwig, Daniels, Callahan, and I have explored can all be extrapolated to the global case, and that although they do not under currently global conditions show that we in the first world have a duty to refrain from claiming expensive medical resources as we are about to die, much less have a duty to die, in order to promote life span equity among the inhabitants of the world, we may well have an obligation to promote international structures of transfer and redistribution of health care savings, against a background of mutual health-related obligations, which would mean that choices concerning dying in the first world could directly affect life prospects in the second and third. If we do this, however, we then will indeed have duties to refrain from claiming resources, or even, according to some of these arguments, a duty to die.

Is this argument a *reductio ad absurdum* of the Hardwig, Battin, and Callahan claims, or a genuine conclusion, although, as Peter Unger would put it, a "highly demanding" one?[25] I would like to predict that with time, and with growing recognition of the interconnectedness of the interests of the various peoples of the globe, including not only our obligations to other peoples but their obligations to us, we will come to see it not as a silly thought-experiment, but as a real challenge to our moral selves. It might seem to be bad news for those of us in the fortunate situations of the developed world, enjoying our long life-spans, but viewed from a Rawlsian/Daniels perspective, it would promote the good of us all by producing a far more just health-related world.

Notes and References

I would like to thank the faculty and graduate student members of the Graduate Philosophy club at the University of Utah for discussion of an earlier draft: Bruce Landesman, Leslie Francis, Troy Booher, Susan Downs, Brad Hunt, Karthik Nadesan, Margaret Plane, Jacqueline Solon, and Ryan Spellecy.

[1]John Hardwig (1997) Is there a duty to die? *The Hastings Center Report* 27(2):34–42 (March–April).

[2]Margaret P. Battin (1987) Age rationing and the just distribution of health care: is there a duty to die? *Ethics* 97(2):317–340. With slight modifications this piece is also reprinted in my volume *The Least Worst Death: Essays in Bioethics on the End of Life*. Oxford University Press, New York, 1994, pp. 58–79, under the title Is there a duty to die? Age rationing and the just distribution of health care.

[3]I would like to thank Ryan Spellecy for his contribution to this question.

[4]Life expectancy data for 1995 from the United Nations Development Program, *Human Development Report 1998* (Oxford University Press, 1998), table 1, pp. 128–130.

[5]John K. Iglehart (1999) The American health care system: expenditures. *N. Engl. J. Med.* 340(1): 70–76.

[6]Ibid.

[7]Jan Narveson, “We Don’t Owe Them a Thing!”paper for the Pacific Division meetings of the American Philosophical Association, April 1999.

[8]Tris Englehardt’s various discussions of the distinction between the unfortunate and the unfair might be extrapolated in this way.

[9]Hardwig, p. 301.

[10]John Rawls (1971) *A Theory of Justice*. Harvard University Press, Cambridge, MA.

[11]Norman Daniels (1988) *Am I My Parents’ Keeper? An Essay on Justice Between the Young and the Old*. Oxford University Press, New York.

[12]Ezekiel J. Emanuel and Margaret P. Battin (1998).What are the potential cost savings from legalizing physician-assisted suicide? *N. Engl. J. Med.* 339(3):167–172.

[13]Daniel Callahan (1998) *False Hopes. Why America’s Quest for Perfect Health Is a Recipe for Failure*. Simon and Schuster, New York.

[14]Daniels, ibid., Why saying no is so hard, in *Am I My Parents' Keeper?*

[15]In the Rawlsian view, this is what would be required for the existence of such obligations. *See* John Rawls, The law of peoples, pp. 41–82 in his *On Human Rights.* Oxford Amnesty Lectures. Basic Books, New York, 1993.

[16]*See* the impassioned analysis by Peter Unger, *Living High and Letting Die: Our Illusion of Innocence.* Oxford, New York 1996, and the work of Peter Singer.

[17]Hardwig, Is there a duty to die? p. 37.

[18]*See* the chapter by Ryan B. Spellecy, Dying for others: family, altruism, and a duty to die, in this volume.

[19]Life insurance would not be invalidated by withholding or withdrawing medical treatment, even if the patient's intention were to die; suicide, under most policies, would invalidate it only within a stipulated period after purchase (in most states, 2 years), and reduce double-indemnity insurance to payment of the face amount. In Oregon, at the moment the only state permitting physician-assisted suicide, the law permitting a physician to provide a lethal prescription to a terminally ill patient at that person's express request stipulates that the act of using it shall not constitute a suicide.

[20]Rawls, The law of peoples, pp. 52,77.

[21]Rawls, The law of peoples, p.75.

[22]Rawls, The law of peoples, p.75.

[23]The author thanks James Humber for this point.

[24]1990 data, *Human Development Report 1998,* tables 12,17.

[25]Unger, *Living High and Letting Die.*

Abstract

John Hardwig argues persuasively that we should regard ourselves as having a duty to die, under a fairly likely range of circumstances. In fact, the very old, when not in good health, characteristically see life as a burden, and need no "duty" to die; they seek only easy means at the appropriate time, and are more concerned to have a right than a duty to utilize those means. The Hardwig thesis applies when our continued life places an excessive burden on those near and dear. Understood in this manner, he is broadly right, as long as we understand this kind of duty as not entailing enforceable requirements, like the rule against murder, but rather as being part of our commitments to the people we would be imposing burdens on. The Hardwig thesis may be accepted, if understood as a broad recommendation to people whose means are limited and who have family and friends who matter to them.

Is There a Duty to Die?

Jan Narveson

A Personal Introduction

A few years ago, my parents, aged 90 and 91, died after long, generally healthy, very active, and useful lives. My father, who died first, loathed and dreaded the very thought of nursing homes. Though he had for some years suffered from an obscure arthritic condition and was bent considerably over, greatly reducing his agility, he had nevertheless been able to make himself useful at home—especially, useful in tending to the needs of my similar-aged mother, both of whose legs had been amputated a few years before. Both of my parents had been active, outgoing, hardworking people all their lives, and both were anxious to remain for as long as possible in the nice home they had built, decades previously. One day my father had a nasty upsurge of his prevailing malady and was rushed to the hospital. From there, he was taken to a nursing home, and while there, he became convinced that he would not be able to return to his beloved home or be of any use to his beloved wife. Therefore, he refused all treatment and food, and died soon thereafter. My mother lived on a few more months. She too went to a nursing home, there being no one who could care for her at home, and it was indeed a miserable place—not for lack of care, sanitation, or activities, but because it was full of people whose lives were seen by them to be basically pointless.

My mother had a mild stroke or two not so long after my father died, and lacking legs as she was, she was in pretty poor shape. In her last days, faced with an operation with only a 10% chance of success, her children near to the scene requested it not be done, acting on what they perceived to be her desires in making this request. She too died soon thereafter. About the same time, my mother-in-law, also aged, developed complications and she too died on refusing treatment for an ailment that although a nuisance, could probably have been cured—enlisting her daughter and a trusted friend to make sure that the medical personnel on hand did as she wanted.

When I told these stories to various of my colleagues, I was interested to find that not just one or two, but nearly every one of them whose parents had died peacefully at advanced age had similar stories to tell. The parent or parents in question had been moved to a nursing home, hated it, and soon, for all practical purposes, died voluntarily, refusing treatment at strategic moments. It was not active suicide, but more like voluntary passive self-inflicted euthanasia. I was impressed by this, and a bit of reading soon persuaded me that there was nothing unusual about either my parents or theirs. When people get very old and are scarcely able to lead active lives, they quite typically not only lose their will to live, but develop a will to die, and they act on it.

Do such people act from a sense of duty? Most of them are on health plans, and part of their motivation may well have been that they were living on other people's money. However, most of it, I think, was simply a sense that continued life was simply not worth bothering with. Not duty, but interest, or an ideal of life is the predominant motive.

My own parents were very religious people, in a protestant way, and this gives rise to a well-known conflict. On the one hand, they believed that death would remove them to a better life; on the other, they were taught that suicide is wrong. I was not near my parents when they died, and what they said about death when I did see them was insufficient to clarify that. However, I conjecture that had they been utterly devoid of religious beliefs, it would have made no difference: they were active, busy, outgoing people, and when life on satisfactory terms to such people is no longer possible, they want out. If duty entered into it—which it may have—it was strictly a supplementary consideration. (My parents, in fact, took care to transfer the family holdings to their eldest child in order to make sure that it would not be devoted to keeping them alive.) Insofar as duty entered the picture, what kept my father alive during his last few years was his sense of duty to his wife, who was very dependent on his care. My father knew that my mother did not want to go to a nursing home any more than he did, and in remaining alive to help her, he undoubtedly provided her with several extra years of reasonably satisfactory life. Her subsequent experience in the nursing home confirmed the wisdom of his assessment of her needs.

However, I am very sure that neither of them had John Hardwig's[1] or Richard Lamm's[2] view. Hardwig says, "there can be a duty to die when one would prefer to live" (35), but if my parents or most of those I know of who died willingly at the end would have preferred to live on, then they would not have done what they did. For them, the difficult thing was life—getting through the next day with at least some sense of accomplishment or pleasure. Dying was easy for them when the time came, because it was precisely what they wanted to do. If the fact that they thereby saved other people a fair amount of money was a factor, it was a minor one.

The Logic of Duty and of a Duty to Die

Can we have a duty to die? Duties, to begin with, apply directly only to actions. The expression "a duty to die" is prima facie a neologism. Therefore, to make the issue clear, we must say that what is meant is a duty that we would have to discharge by doing what results in our deaths. There are a number of ways: suicide, of course, but also, as in the case of my father, refraining from taking actions that would prevent death. For example, we might be on an elaborate life-support system and ask that it be switched off. We could refuse further treatment. Or, somebody is about to kill us, and we would be able to take preventive action, which would probably succeed, but we do not take that action.

However you take it, though, a duty to die is on the face of it bizarre. We may think that suicide is our best option, and do so out of self-interest, but how would we have a moral duty to do this? To whom and why? It is easy enough to dream up special cases. For that matter, we do not need to dream: the case of the spy who has just been caught and whose duty is to bite on the cyanide capsule concealed in his teeth rather than divulge information to the enemy has actually happened more than once.

In general, these are scenarios in which we owe somebody a duty such that in order to discharge it, we must do what causes our death, but is it possible for there also to be a duty to die, as such? It is hard to see how; the idea seems to make little sense. Would this be a duty to oneself? That is an odd category to start with. When we say that we "owe it to ourselves" that we do such-and-such, that is a claim that in some quite fundamental way is nobody else's business. Moral duties proper, however, are always somebody else's—indeed, everybody else's—business. Society may compel us to what is our moral duty.

I might owe it to myself to practice the violin more, to take a vacation, or to stop eating some food or other that I like very much, but that is doing something awful to me. However, if the category of duty applies at all, it stems from my interest in living and living well. That is not moral duty as such, in the narrow and useful sense of the term in which our duties are what we owe in consequence of our social condition. Such duties are essentially to others, even though the considerations that give rise to them are, no doubt, anchored in our interests as humans. However, how could a duty to die fit in with that general idea?

Two interesting possibilities come to mind. One I will dismiss immediately: it is that we simply "belong to" the community —

fascist morality. I assume that all readers, including Hardwig, really want to reject that. Whatever there is to his anti-individualism, I trust it is not that.

Owing It to Oneself to Die

The other is this: we can imagine someone having a certain view about life that included a notion of "completeness." Now this person might think that his or her life is complete at a certain point, and that at just that point, he or she should, if nature is not doing it for him or her, bring about his or her own end. This person acts in the fullness of time, as it were—to make his or her life complete, in the right way. But again, this is not morals, but rather, a humanistic vision, even an esthetic vision of life. I take it as obvious that nothing of that kind can be a requirement that we may impose on our fellows. Instead, it is something we can live by ourselves and recommend to others for their consideration. However, I take it to be axiomatic that my view of how to live has no authority over you. Any requirements of the kind morals aim to impose will have to be based on something quite different. They will have to be based on our general relations to each other as persons interacting in society. The thesis that there is a duty to die is one about our duties to others, duties that turn out, owing to one circumstance or another, to have the unhappy upshot that in order to fulfill them, I have to elect death. What could do that?

Dying that Others Might Live

In general, I should think, any such duties will have the structure that I am to die in order that certain others might live, or continue to live at something like the level they previously enjoyed. However, that does not sufficiently explain things, either. For we do not in general have a duty to die that others might live:

indeed, we do not in general have a duty to help others to live, well or at all. If you will die unless I do *X*, I still do not in general have a duty to do *X*, just like that. But it may have been well for me, at some time, to make an agreement with others that is to the advantage of all of some set of people, and that makes it a duty of all signatories to move over, as it were, in certain contingencies, and surrender their place to some other person.

That is far from inconceivable. Perhaps I am here now only because someone else also made this agreement, and kept it—and now it is my turn. In wartime situations, my fellow platoon members and I must sometimes do very dangerous things, without which we all or most of us will be killed, and it can be my turn now: I am the only one, perhaps, who is in a position to leap up, drawing the enemy's fire, but enabling the other to escape. If so, that is my soldierly duty, and it might amount to a duty to die, or close enough to do.

Civilian analogs of that are not as easy to come by—fortunately! However, they are perhaps not impossible. The most plausible scenarios, I suppose, do lie in the province of the medical. I might have signed into a kidney-sharing arrangement, perhaps: I am alive now only because so-and-so gave me his kidney; but he did so on condition that I will in turn give mine to somebody else. It is not so easy to flesh out that arrangement, but I do not see why it could not, in principle, be done. It would, of course, take some serious soul-searching to sign into such an arrangement, and it is easy to imagine that someone who had done so, and whose turn came up unexpectedly, might balk at doing his or her duty. Our reluctance to enforce such a duty is unsurprising, but that it would be his or her duty seems clear enough.

Hardwig emphasizes loved ones—family, especially, and friends, and we may surely agree that our duties to such persons are not a matter of a straightforward agreement. On the other hand, they are matters of interest, and families are not sources of absolute moral imperatives. We can, and occasionally do, cut off children with a penny, we leave spouses, parents, and children. Sometimes people do this when they ought not to have, no doubt, but in general, they have the right to do this. The state should not be in the business of destroying families, as it is widely accused nowadays of doing, but neither is it in the business of shoring them up by main force. My conclusion is that duties to families and friends are also a function of what amount to agreements, though usually unstated and not specifically entered into at particular times. We stay with our loved ones because we love them, or because we have pride or a desire to enhance our gene pools, and perhaps for other reasons. However, we do not owe anything to children, parents, sisters, and others simply because of a given genetic relationship. Certainly, our friends must be chosen. We stick with them because we love them, as well, often, as because they have done much for us. In the case of families, the ties are there, but they are accepted and woven into our lives by emotion and will, not by molecular force.

Duty as Onerous

In all this, of course, I assume that death appears to the agent as an evil. Some religions promise immortal life after our earthly sojourn is over. Such religions might make death out to be an attractive idea, really just a crossing of the threshold to a much superior sort of life. One might characterize those views as really denying that people die at all, strictly speaking. At any rate, in the following, I assume that such views are not held by the reader; for those who do hold them, there is little point in reading on. For them, other discussions are in order.

Hardwig concluded his envisaged list of conditions in which there is a duty to die with the thought, "Finally, there can be a duty to die when one would prefer to live"(35). However, I would suggest that this is not just another item on a list: it is, rather, a defining condition of what it is to have such a duty. Anything short of that is not a duty, but a means of self-improvement, or living the best life one can manage, but if I must die even though I do not want to, and would not need to, given available alternatives, then indeed we are speaking of duty, properly speaking. That applies to the soldier who smothers the grenade, the member of an insurance group whose turn has come up, and the intelligence agent caught and facing torture. However, old people such as my parents do not meet this condition. They prefer death—comfortable, if possible, yes—but mainly, please, soon.

Dying and Distributive Justice

Returning to medical scenarios, let us first address ourselves to a fairly widespread view that would seem to imply that the duty to die is much more likely than we might have thought. The framing assumption is that continued life for some number of people, N, requires use of a scarce medical technology of which there is enough available only to accommodate M, N being greater than M. How, then—so the question is framed—are we to distribute this resource fairly? Should we give everyone an M/N crack at it, randomizing by some good method? Or should we take the number of years of life expectancy otherwise remaining into account? Or what? Whatever, the thesis is that the losers now have, indeed, the duty to die. Actually, since the required procedure will not be available in their case, they will not be able to do anything about it, but we can still make this into a duty to die by supposing that the relevant M people initiate the randomizing procedure themselves, for instance.

However, there is an assumption here that may and should be disputed. The assumption is that all of these people, just by virtue of needing this procedure in order to continue living, are thereby entitled to it. We should surely reject that assumption. People who save other people's lives are not in general doing something that their beneficiaries are entitled to. On the contrary, they are doing them a very great favor, for which, if the beneficiaries get it for nothing, they ought to be very thankful.

In the normal case, of course, they will not get it for nothing. Those who provide the benefit will be paid, probably quite handsomely, for their work, and those who received it will either have bought the procedure and be entitled to it for that reason only, or they will have paid into some scheme, or perhaps, as in Canada, into an involuntary government-arranged scheme, which results in their having (or not) the procedure in question.

Which among these is right? I should think it pretty obvious that the right one is that wherein applicants for the procedure pay for it, as a service that is worth at least that much to them. The price might be very high, in which case the procedure will not be worth it to some people. Even if they could afford it, they might do better keeping the money in the bank for the benefit of their widows or children, say, or giving it to some cause they consider more important than this. Also, if the price is simply beyond their means, then unless somebody else decides to buy it for them, they will not get it and they will die sooner, but this will not be because it is their duty to die. It will be simply because they can not afford to continue living.

Old-Age Egalitarianism

Some at least profess to think otherwise. They think that we must all be willing to share medical procedures with all and sundry, on an equal basis, however expensive, as long as they are necessary for continued life. It must be asked why they think thus. We assume, of course, that we are not referring to cases in which we ourselves have caused the victim's critical condition and so are responsible for trying to rectify his or her situation, even if this would cost our own life. Rather, we are talking about simple cases of limited supplies of what turn out to be life-saving devices or procedures. Why do individuals think they are special?

One way of saying that they are not special is that of the out-and-out egalitarian, who asserts that we ought in general to share all good things equally. That is a view not widely shared—especially in practice—and I will assume that it is not held, or even seriously entertained, by the reader. But if it is not assumed, and yet it is thought that medical services are things we owe people, then the question needs to be addressed why these particular services should be treated so differently from the other good things in life that the activities of others have created or supplied. There would seem to be an assumption that things like medical life-saving procedures are, from the moral point of view, qualitatively different from other services.

But are they? I find that dubious. What makes it especially so is that all sorts of ordinary consumer goods promote life in one way or another. The new car that enabled you to get to the hospital with your seriously injured daughter just in time is instrumental in prolonging her life, and is the wholesome food we purchase with our middle-class incomes.

More important, perhaps, is that when we buy anything, we hope thereby to make our lives go better in some way. Sometimes this translates into life expectancy. More often, however, it translates into quality of life. We think we live a better life doing *X* than doing *Y*: going to the opera, say, instead of much more economically staying home and watching TV or reading the newspaper. Also at still other times, it translates into quality at the expense of quantity.

Are quality and quantity incomparable? Not at all. The rational smoker can decide, quite consciously, to take his or her chances on an expectedly shorter but pleasanter life of the smoker rather than the less pleasant but longer life of the nonsmoker. Racing car drivers, mountain-climbers, and any number of others have surely concluded that the superior thrillingness or richness of their chosen lives outweighs their likely shortening.

Individualism

John Hardwig accuses those with the stubbornness to continue insisting that one's life is, after all, one's own, of harboring an "individualistic fantasy" (35,36). According to it, he says, we "imagine that lives are separate and unconnected, or that they could be so if we chose" (35), but I suggest that his argument involves a muddle, worthy, perhaps, of another label: the Separatist Fallacy. According to this fallacy, individualism entails unconnectedness—that we do not affect each other. In short, Hardwig ascribes to defenders of individualism the view that we are all really Robinson Crusoe, only without Friday. But surely nobody has ever thought any such thing. When we say that people are distinct individuals, we mean nothing of the sort. What we do mean is that if something happens to person *A*, whoever *A* may be, it does not necessarily, as a matter of logic, have any particular effect on person *B*, whoever *B* may be. However, the fact that what happens to me does not necessarily affect you certainly does not imply that it does not in fact affect you. If my wife died, that would affect me plenty, but the effect would not be a matter of logic.

How does the "fantasy" bear on the present question? According to Hardwig, if the sort of individualism I and most of us subscribe to were true, then "the relevant questions when making treatment decisions would be precisely those we ask: What will benefit the patient? Who can best decide that? The pivotal issue would always be simply whether the patient wants to live like this and whether she would consider herself better off dead." Also, he adds, "'Whose life is it, anyway?' we ask rhetorically" (35).

Well, some of us do not see this as rhetoric at all. Frankly, we think that that is the central question. Sometimes there are questions of that person's duties to others, indeed, but as Hardwig suggests at the outset, we suppose that those cases are comparatively rare. If he thinks not, then why not?

There is an answer. He mentions our "deeply interwoven lives" (36), but those with whom our lives are deeply interwoven are different people in each case. Moreover, we really do have a choice whether to associate with those people or not—whether to remain deeply interwoven with them if we already are, or to get deeply involved with them if we are not as yet. At no point does logic dictate that we shall be deeply woven with person *X*, or *Y*, or *Z*, and certainly not with all of *X*, *Y*, and *Z*. Some few people may want to do something like that, but most of us do not, thanks very much.

Hardwig's view seems to be that we are all in a huge medical lifeboat: if person *A* gets some treatment, then there is some person *B* who does not. And in order to make that fact, insofar as it is one, decisive in sticking us with a duty to die, or something like that, we need also to assume the very premise I have objected to above: that medical services, intrinsically, must be shared with all. I reject that, and so does he, really, as does everybody we know. We do not have any duty to do any such thing. Sometimes we are entitled to a certain medical service, and sometimes not. When we are not, no one has a duty to give it to us or share it with us.

Therefore, we can agree with him that we ought not to make ourselves a burden to people, and that prolonging our lives will sometimes bring us into that situation. When that happens, we should consider whether it is worthwhile. It may not be, but our question here is whether it is the duty of those who bear this burden to bear it or as much of it as they do. They may themselves be acting out of a sense of duty, and that sense might be justified, but it might not.

Socialized Death?

Some people would respond to Hardwig's argument by proposing that we ought to broaden the base, as we do in Canada, with its socialized health care system. He himself suggests (40) that the duty to die might be virtually eliminated by our society "providing for the debilitated, the chronically ill, and the elderly." Aged persons in Canada are not a financial burden on their families, indeed, but instead they are a financial burden on everyone in the whole country. As the population ages, this burden gets larger and larger. Tax burdens in Canada are much higher than in the United States, and in fact are among the heaviest in the world. To my mind, it is very questionable whether we get our money's worth, either from the health care part of this burden (which is enormous) or from the rest of it (also enormous). The American system has the advantage that the burdens of caring for the aged are really felt by the people who are closest to those aged or incapacitated people. The very old who are in such situations will correctly perceive that they are being a great burden on certain particular other people—people they care about. A socialized health care policy masks that, and in the process increases the per capita costs of care tremendously. (The United States is, by the way, far more socialized than he may realize. In fact, every individual in the United States may receive unlimited care, once his or her own resources run out. This, too, has already hugely increased per capita medical costs.)

Hardwig discusses three objections to his view. I have no sympathy with two of the ones he mentions: the supposed higher duty to stay alive, and the dignity of the person, and would second much of what he says about those, adding that it seems to me that to be incompetent in the way that the very aged tend to become is a lot less dignified than saying "Enough!" and pulling the relevant plugs. His third objection appeals to the burden placed on the person who is made to feel that he or she has a duty to die, on top of what he or she is already suffering. His immediate reply to this is certainly correct: it is not obvious that the burden on me of facing up to my near-future mortality is as great as the burden on those who are paying the bills to stave off that future a little longer, but of course that matters to me only insofar as I am sympathetic to those facing the burdens.

Hardwig does not discuss the important, but tangential subject of why the costs of late-term care are so high. I suppose that like most writers, he simply assumes that modern medicine is inherently expensive and lets it go at that. A more careful look, however, would probably show that it is the actions of modern government, and not modern medical researchers, that add most to the bills. However, that is a side issue in this discussion, and I will not pursue it further.

Personal Duties

I agree fully with him that we ought not to prolong our own lives to the ruination of people we love, or for that matter, of people we do not love. In part, this can be headed off by improved insurance procedures and more efficient care systems—none of which will be provided by your government. However, insofar as the situation is as he says, I do not see how any self-respecting elderly person could, or would, dispute Hardwig's general claim here. I would dispute that those persons have a "duty to die" in the same sense that they have a duty, for instance, not to kill, but surely Hardwig would agree with me on that. Still, they ought not to insist on their care-bearers ruining themselves to sustain their lives, and not only ought they not to insist on it, but they should not let them do that. I agree with that, too, but here "duty" is being used merely as "the noun of 'ought.'" However, that is not its maximal sense. The maximal sense obtains when a duty is enforceable, by the community generally,[3] and I do not think Hardwig does mean that the community generally should be able to enforce this particular duty.

In countries like Canada, the socialized medical services are under very heavy financial pressure. Our government has responded to these in part by withdrawing services from the list of available ones. If you need a multiple-bypass heart operation, you may not be able to get it—thousands of Canadians go to Detroit and other border cities to get it at their own expense, but those of their fellows who can not afford it are just out of luck. Under a system like ours, your life is literally in the hands of the government. I do not know whether Hardwig thinks that is okay, but I do not. Your life should instead be in the hands of whatever association you contracted into to manage it in the relevant respects, or were born into and do not want to opt out of, and what happens to it will be a function of your specific agreements or specific sense of obligation to the others. Beyond that, what happens to it calls for individual decision by you, in the light of all relevant factors. Insofar as one of those relevant factors is that you are imposing immense costs on people who matter to you, you have good reason to cease living, and it is a reasonable use of the term "duty" to describe this by saying that you have, in those circumstances, the duty to die. However, I deny that this is something that can be imposed on you by the community in general or by any government in particular—and certainly not by your loved ones.

Hardwig goes too far, however, when he suggests that "to have reached the age of, say, seventy-five or eighty years without being ready to die is itself a moral failing..." (39). Well, "ready" how? Plenty of people reach such an age with no intention of dying any time soon and every intention to fight death to the last. They may not be able to afford to carry on that fight effectively, but they will fight if they can. Insofar as they can do so without imposing burdens on unwilling others, that is surely their call and their right.

To his credit, Hardwig also addresses the difficult case of the incompetent. It is not all that different, really, for old people, just as such, need not impose burdens on others. An 80-year-old in perfect health eats not only no more, but rather less, than you or I; he or she is likely content to walk in the park daily, to read, to watch TV, and talk with his or her friends. These are not inherently expensive activities, although carrying them on in a state-subsidized nursing home may make them so. It is the less than full competence of the aged that makes them expensive, but of course Hardwig means those whose minds are such that they can not appreciate notions of duty and the like. Here he proposes that "I can make no sense of the claim that someone has a duty to die if the person has never been able to understand moral obligation at all"(39). Fair enough. However, the rest of us can address the question whether we have the duty to maintain that person in life, at our expense, and there I think Hardwig ought to say that the answer is that we do not. We can decide that the cost is too great, and allow this unfortunate individual to die—though it will and should be an uncomfortable decision.

At the end of his paper, Hardwig suggests that "we fear death too much. Our fear of death has led to a massive assault on it … We do not even ask about meaning in death, so busy are we with trying to postpone it" (40). I agree with him about that, but I also think—and here he might well agree with me—that the meaning of death, which is surely a function of the meaning of life, is personal and not something that we should strike a Senate Committee to make decisions about, with policies forthcoming based on our results.

That brings us back to the subject of individualism. All duties are social: duty is social, but it is social in being the result of interrelations among individuals, and their individual decisions and judgments are relevant to its content. In particular, duties are a function of what we have agreed to—and what we have not. The sense in which there is no "duty to die" as such is that this duty cannot be simply imposed by society at large, but there is no duty to maintain life, unlimitedly, either. Almost all, I think, of Hardwig's thesis is accounted for by that observation. If he meant more by it, however, then I think we should not accept it.

Notes and References

[1] John Hardwig (1997) Is there a duty to die? Hastings Center Report 27 no. 2, 34–42. Numbers in my text refer to page numbers in that article.

[2] I have no separate reference for Lamm; Hardwig cites him as "claiming that old people had a duty to die" in his opening sentence, op. cit.

[3] J. O. Urmson, Saints and heroes, splendidly marshals the case for a more subtle vocabulary. The essay is originally in A. I. Melden, ed. (1958) *Essays in Moral Philosophy.* University of Washington Press, Seattle, Washington.

Abstract

The claim that we might have a duty to die, particularly when we are old or very ill, seemed at first shocking and harsh. A set of hypothetical cases is examined to show that the claim is morally defensible under certain circumstances. A generic conception of moral duty is described. Two broad categories of moral reasons, those based on considerations of beneficence and those based on ideas about moral principles or virtues apart from beneficence, are studied. Both are shown to provide plausible foundations for the claim that in some situations a person has a duty to die. A summary of factors playing into the duty is given. Finally, practical difficulties that would arise from explicit recognition of the duty are examined along with theoretical complications. It is concluded that the problem, although it clearly causes discomfort, cannot be avoided.

Do We Have a Duty to Die?

Marilyn Bennett

Defining the Problem

The suggestion that there might be a duty to die strikes many as at least callous, perhaps outrageous. A casual remark by Colorado governor Richard Lamm that old, ill people had this duty provoked a furor of response that still reverberates.[1] Lamm responded with an attempt to explain what he had meant by saying, as he later claimed he had, that "we all have this duty".[2] A modest quantity of philosophical literature has examined the problem directly.[3] It appears with some frequency as a secondary issue in discussions of physician-assisted suicide and distribution of health care resources.

There is an immediate conceptual oddity involved here. We tend to think of dying as something that happens to us rather than as an action we might choose, more like being born than paying taxes. Then, too, the assertion of a duty sounds more like an insult—"Why can't they just stop breathing?" or "Drop dead"—than a reasoned moral position. It seems to imply that people can become so tiresome, expensive, or otherwise burdensome that they ought to take steps to eliminate themselves from our presence. Too bad, the idea would go, the rest of us are not in a position to eliminate you, but surely you must see how wrong it is of you to go on requiring our attention, using up our resources, and generally making a nuisance for the rest of us.

What precisely makes this suggestion so offensive? One factor is our ordinarily deep commitment to preserving life. A person's life might be thought of as his or her most valuable possession and the condition of his or her having anything else at all, at least anything of worldly value. Certainly, he or she has a fundamental right to life, one that could not be lost without doing something so horrible that it deserves death as punishment. This right is seen as an autonomy right, so that decisions about how to use it belong strictly to the owner of the life. [4] Another factor may be an evolving liberal egalitarianism that leads us to include increasing numbers in our circle of equally valued fellows. We strive steadily to eliminate arbitrary discriminations. The suggestion that some people or classes of people ought to die reminds us of past moral errors, and this may be what the idea of a duty to die amounts to: It may mean that some people's lives should not be valued with the rest—that our conditions even while we are innocent can fall below some threshold a group agrees to and thus make our elimination desirable. The very old, the chronically, severely, and terminally sick, and the constitutionally defective become victims of the very kind of prejudice and discrimination that we have worked so hard to overcome. Our outrage at the suggestion that they should do away with themselves can serve as a source of pride.

Does a fundamental right to life mean that we are obligated to preserve life regardless of the costs? Certainly, we can find it fitting to break many moral rules in order to save lives. Most of us would steal, lie, or cheat to prevent an unwanted death. However, although a contrary duty goes against the stream, we do in fact require some people to sacrifice their lives for goals other than the saving of greater numbers of lives. Soldiers and police officers can have dangerous duties so stringent that they may not rightly fail in them even to protect their lives. Several rescuers may be required to risk their lives to try to save just one endangered person. It will be objected that these people do not aim at their own deaths in carrying out their duties even when they are virtually sure to die. To meet this objection, it will be useful to consider a few idiosyncratic, perhaps implausible examples to show that it is conceivable that someone might specifically have a duty to die. These are chosen to show a range of possible groundings for such a duty.

The Restricted Trust

A very old person has led a rich and fulfilling life thanks in part to an inheritance that provided him a generous sum each year for the duration of his life and called for the remainder of the estate to be distributed to his heirs upon his death. His health has now deteriorated to the point where he experiences minimal pleasure and much trouble. His young heirs have valuable projects in mind, which they now lack funds to carry out. The terms of the trust prevent him from giving away more than his allowance in any given year. In this case, we might think that although no one has done anything to deserve death, everyone in the family would be better off if the old man died.

The Captive Maiden

A young woman has been captured by members of an indigenous tribe who will use her horribly. Her loved ones (along with the one remaining Mohican) watch in helpless anguish as she is dragged up a mountain path on a rope. She sees an opportunity to break free and throw herself over a cliff at a place where the family would observe this.

Here, it would be much better for the young woman and her family if she died, though the captors will suffer a loss. The case is complicated by her being the victim of a wrong, but none of the people who are being harmed is in a position to right it.

The Terrorists' Pawn

An adventurous and charismatic traveler has persuaded friends to accompany him into a dangerous area where the party has been captured by terrorists. He could have learned of the danger, but did not take the trouble. His friends could not have made successful inquiries independently. The terrorists announce that they will murder one hostage to show that they are serious; they will accept a volunteer. Here, even if the adventurer is the most likely among the hostages to be able to negotiate the group's ultimate safety, it seems most just that he should volunteer to die since he has led the others into the danger.

Casting Lots for Air

Two explorers find themselves trapped in a cave containing a fixed amount of oxygen. They calculate that if both go on breathing, neither will live until the rescue party arrives. However, there should be enough oxygen for one. They agree to draw lots to see which one wins the oxygen and which one will discontinue breathing.

Suppose that, if the situation is anyone's fault, the fault is shared equally. At least one person will be made substantially better off if the other dies, whereas neither will benefit if both die. Since each person has an equal chance of winning the lottery, the loser cannot complain of injustice.

The four cases will not be found equally persuasive by everyone. It seems very likely that at least one of them will be found persuasive. If so, then either considerations of well-being or considerations of justice or both can lead us to admit that a person could have a duty to die.

Let us first consider what would reasonably be meant by "a duty to die." We can consider this in a generic way, apart from particular moral theories, by thinking of the question like this: what would be meant by saying that a person has a strong, perhaps overriding, moral obligation to act so as to bring about his or her own death? The generic version will assume the following points:

1. To say that one has an actual duty to do a certain thing is to suppose that it is an available option. It must be something the person is capable of doing and there must not be special circumstances that prevent him or her from exercising this capacity. Where circumstances interfere with

someone's ability to act, we may say that he or she was excused from doing his or her duty because he or she was prevented or we may find that he or she had some other duty under the circumstances.

2. The act in question must be one that an agent may intentionally choose. Since we will all die eventually regardless of what we choose, the force of a claim that we have this duty must be directed at actions we can ourselves take, presumably to allow or produce death earlier than it would happen if we did not act.
3. A duty to die might be carried out passively, by forgoing measures to prolong life, or actively by taking lethal measures.
4. A duty may or may not correlate with a right on someone's part to demand its performance. A duty might correspond to a specific right held by a specific person (a duty to pay a debt) or to a more general right held by unspecified people (a duty to avoid toxic pollution). Duties need not correspond to rights (a duty to act kindly may not be grounded by anyone's right to your kindness).[5]
5. If someone has a duty to perform a certain act, the duty is present regardless of his or her desire to perform it. If he or she wants to do what is in any case his or her duty, so much the better.

Once this minimal set of generic-duty conditions is specified, a distinct set of related problems remains.

6. We will still need to determine whether a particular duty is enforceable. There are some duties that will be forced on us if we fail to honor them voluntarily (paying taxes, driving sober). Other duties lack this feature.
7. We will still need to determine what appropriately follows from a person's choice not to do his or her duty. For instance, if a person refuses to carry out a duty to die and this duty is not enforceable (refusal does not result in our

having a right to kill him or her), a question remains about whether that person has a right to continue using health care resources.

8. The wisdom of openly acknowledging a duty considering the consequences of anticipated abuse or misconstruing that duty in a particular social climate may be a relevant concern.

Beneficence and a Duty to Die

One quite plausible way of determining our moral obligation involves considering which actions of ours could be expected to produce the best states of affairs. The classical utilitarians have provided perhaps the most familiar theoretical framework for this view, though there are certainly well-known and respectable variations. We can characterize this position generally by saying that its adherents hold that morally correct actions aim at beneficence: our goal should be to make some set of subjects (perhaps the world) better off, to act in their best interests.

On this approach to morality, it is misleading to think of our moral obligations as distinct from the burdens involved in carrying them out. We determine our moral obligations by weighing the benefits and burdens connected with our options. If the negative consequences of a particular action are too great, it is simply not the correct action. It is not seen as "a duty we cannot afford to honor," but rather as not being the morally correct choice. If someone using the beneficence approach says, "Naturally we ought to make it possible for all who want to go on living to be given every opportunity, but sadly, resources are limited," this will mean something like, "We would be happy to provide resources if they were more abundant, and needy people and their loved ones would often be spared sadness, but the fact is that we do not find this to be the most efficient use of our present resources and therefore we ought to be doing just what we now do."

The substantive problems in using this approach arise when we try to weigh and distribute burdens and benefits. When is continued living a benefit? Is death a great harm? If 20 people are relieved of great pain at the cost of killing one person, is the overall state of affairs better than if the one goes on living and the 20 go on suffering? If your dollar would feed a family for a week, will things turn out better overall if you spend it on a newspaper rather than sending the dollar to them? A number of contemporary discussions of such problems are relevant here. Murphy gives a particularly helpful review of beneficence-based theories that includes a discussion of fair distribution of burdens.[6] Others have focused specifically on weighing continued life against competing values.[7]

Most seem now to agree that there are circumstances to which anyone might prefer death; many now hold that this preference can be rationally defended. A compelling summary of reasons appears in a very recent study of suicide:

> Studies of suicide among the elderly do find correlations between physical debility and increased incidence of suicide. There is some evidence that chronic dyspnea, for example, is a risk factor both for depression and for suicide among the elderly. Physical debility can lead to significant dependence on others to meet basic needs of life: shopping for food; cooking; transportation to social occasions, church, or even to doctor's appointments, cleaning, and the ability to continue to live at home. Loss of bodily function, such as incontinence, can be the source of humiliation and self-disgust. Declining vision or hearing can make cherished pastimes such as reading, sewing, or listening to music no longer possible. Physical losses thus may result in other losses, of pleasures or independent living... Beyond the physical, elders are more likely to be subject to other kinds of losses. They are more likely than other age cohorts to experience the deaths of family members or friends. Despite the legal end of mandatory retirement, age at retirement has been falling and the elderly are more likely to experience job loss and resulting loss of activity and identity. Remaining family members may be physically or emotionally remote, particularly

> in the face of apparently increasing needs and demands. Even the loss of pets, due to death or the inability to care for them physically or in reduced surroundings, is more significant in the elderly. Moves—for family proximity, for ease of physical maintenance, or for nursing care—represent the loss of perhaps a lifetime of familiar and cherished surroundings. All of these losses appear to increase suicide risks among the elderly... One vulnerability the elderly do not suffer from disproportionately as a group is poverty.[8]

See also Prado[9] for a discussion of factors leading people to choose suicide before these losses are actually suffered. An earlier volume edited by Battin and Mayo explored the rationality of suicide in some detail.[10]

A full beneficence calculation requires consideration of more than whether a particular individual would be better off dead. Hardwig details the considerable costs to particular others of caring for an ill or dependent person in a society where it is not common for anyone to be at home and available to help with daily needs for care and companionship, transportation to appointments, and so on. He describes a situation in which a middle-aged daughter has sacrificed her job, her career, and her savings to provide for her mother's needs. Clearly, there is a chain of costs involved here; the daughter who has made these sacrifices will become a needy person herself, and others will be required to make sacrifices to meet her needs later on.[11]

It is, of course, not only family members and specific others who bear burdens in caring for the sick. Of particular interest is a report on monetary costs involved in the last six months of life. Figures from the Health Care Financing Administration for 1994 indicated that about 21% of Medicare spending for the year went to the approximate 6% of beneficiaries who died and that expenses for the final year were about four times as much as for earlier years. It is supposed that at least an equal amount is spent during a person's last year from other sources. The figures suggest that earlier deaths would substantially free health care funds and other resources for use by others.[12]

A critical question arises here. Does the beneficence-oriented agent have any reason to believe that resources he or she arranges to save (by bringing about his or her own earlier death or by using his or her influence to bring about others') will in fact be spent on those who have more to gain from receiving them? Sometimes, as in cases of organ scarcity, treatment of one person affects the treatment of others. In a "zero-sum" system, whatever is spent for one patient means exactly that much less is available for others. It is pointed out that most patients in the United States are in an open-ended system rather than a zero-sum one. Spending public funds on them does not directly affect the care available for others. Where this is the case, the beneficence calculation will involve a relatively difficult assessment of the indirect effects of health care spending as well as nonmonetary considerations.

Our ideas about fairness and entitlement compound the beneficence problem. If we see certain resources as legitimately belonging to an individual (his or her own money, for example, or funds earmarked for people in his or her situation), our ability to think of a more beneficent way of spending the resources than the one he or she has in mind has less moral weight than if the resources were not clearly (or not at all) his or hers. Respecting entitlements carries its own set of benefits and burdens, which remain in place as we make the calculations under discussion. Various aspects of desert may also come into play. A 70-year-old person may be just getting free of a lifetime of burdens and sacrifices for others when he learns that his remaining time will be limited and costly. Will he and we believe that the world is made better if he makes a fast exit? Perhaps not; he and we may find that rewarding his loyalty and endurance with some sacrifices of our own would produce a higher sum of benefits and burdens.

Would excessive or discriminatory burdens fall on the sick and dying if we expected them to die earlier so that others could be spared? There is indeed a discomfort produced by the thought of trading off lives for money or time with children or better public parks. Are these things really commensurate? Are some sacrifices just too much to ask?[13]

A Duty to Die on Nonconsequentialist Grounds

Duties are not necessarily grounded by their connection with bringing about desirable states of affairs. Often, actions are seen as morally required because of their character of respecting rights, honoring ideals of justice, or otherwise recognizing moral principles or displaying virtues. Those who hold such principles (unless they hold beneficence as the only one) will act on them even when they do not maximize beneficence. Someone who holds a principle of honesty is likely to ground his or her truthfulness in respect for others' right not to be lied to rather than in a belief that honesty always produces the happiest consequences.

Are there nonconsequentialist principles that could yield a duty to die? As we saw earlier, we can have duties to act in ways that will almost certainly result in our death. The dangerous duties of soldiers and police officers bind them even when a beneficence calculation would show that failing to do these duties would produce better results. The soldier is not excused from his or her post by an argument that more good would come from abandoning it. A parent who abandons an existing child to danger on the grounds that if he or she survives he or she can have more than enough new children to offset the loss is not likely to be found to have carried out his or her parental duty.

Again, in these cases, it is not death that is called for, but actions that are fairly certain to result in death; we hope that by some lucky chance the person whose duty is to defend a position or rescue others may survive the performance of the risky duty. It remains to be seen whether there are recognizable duties specifically to die.

Following a distinction offered by Philippa Foot, we might consider duties as falling into two broad categories. Foot's "duties of justice" are those that require us to do what is owed to others because of their rights to claim these from us. Her "duties of beneficence" require us to act for others' benefit in cases where they cannot claim such benefits as their right. She suggests a rough correspondence between duties of justice and "negative duties" to refrain from interfering with others and between duties of beneficence and "positive duties" to act in their favor.[14]

Examples of arguments for a duty to die grounded in duties of justice can be shown. Consider the case described above in which a foolhardy traveler had led friends to be captured by terrorists. If he indeed has a duty to volunteer when their captors call for a victim, this duty seems most plausibly to arise from a debt he owes to his friends. If one of the group must die, it is only fair that he should be the one. Alternatively, consider a case in which someone has created a situation in which all concerned would be better off if she died, not because of exotic risks, but because of her terrible conduct in the past. Her continued existence stands in the way of family healing, her spouse's remarriage, productive use of the family's resources, and so on.

In both cases, it is plausible that someone owes others his or her own death under the circumstances. This duty flows from a demand of justice that we try to put right what we have damaged or wronged. The idea is not that the person whose death is called for is less valuable than others or has less right to life; it is that a duty of reparation can in some circumstances be strong enough to override a right to life.

A more general duty to die can arise in situations of scarcity of resources and social contract. We could easily understand if, in an artificial colony, a bargain were struck in advance about criteria for deciding who would be sustained in it if resources became so limited that not all could be sustained. Such a scarcity of the necessities of life (the "lifeboat case") gives rise to a familiar problem of justice in distributing resources that are seen as belonging to a cooperative group rather than as individual property. Covering more realistic instances, a great quantity of bioethics literature deals with "fair equality of opportunity" or the "fair-innings argument." Here it is common to see the suggestion that a person might use up his or her fair share of resources, possibly by living to an adequately ripe age to have experienced his or her share of good things, having had extraordinary needs met for a sufficient time, or having such weighty needs as to represent a "black hole" into which resources would be wasted.[15]

In recent years it has been argued that we have a right to preserve our own integrity by preventing our "degradation" or disintegration (where "disintegration" literally amounts to the loss of an integral self). Ronald Dworkin claims that we ought to give moral weight to honoring our intrinsically valuable genuine selves. A possible consequence of his view would be that we have a duty to put a stop to relentless deterioration or essential adulteration of the self.[16]

Might we have, in addition to a duty not to try to take more than our fair share, a duty to temper (as in intentionally lowering) our expectations about our entitlements? Many now comment on our steadily increasing demands for health care. As more interventions become possible, we seem automatically to assume that they ought to and will be used for us. However, it is equally often pointed out that in other respectable nations, people do well with much less. Such a duty is presently harder to argue for than the others reviewed so far. It is counterbalanced by a well-recognized duty to maintain expectations of fair and equal treatment and avoid self-effacing, victim-like behavior.[17]

We can easily add to the list developed so far. Some may recognize duties of honor. The captain no longer goes down with the ship on maritime principle alone, but might do so out of loyalty to others who are being lost. Virtues of generosity and gratitude might come into play, as might aesthetic and spiritual values that can be recognized through one's choosing an earlier death.

Some Plausible Suggestions About a Duty to Die

How can we integrate these theoretical considerations in evaluating the strength of a duty to die? John Hardwig offers a neat summary of suggestions:

1. There is more duty to die when prolonging your life will impose greater burdens—emotional burdens, caregiving, disruption of life plans, and yes, financial hardship—on your family and loved ones. This is the fundamental insight underlying a duty to die.
2. There is a greater duty to die if your loved ones' lives have already been difficult or impoverished (not just financially)— if they have had only a small share of the good things that life has to offer.
3. There is more duty to die to the extent that your loved ones have already made great contributions— perhaps even sacrifices— to make your life a good one, especially if you have not made similar sacrifices for their well-being.
4. There is more duty to die to the extent that you have already lived a full and rich life. You have already had a full share of the good things life offers.
5. Even if you have not lived a rich and full life, there is more duty to die as you grow older. As we become older, there is a diminishing chance that we will be able to make the changes that would now be required to turn our lives around. As we age, we will also be giving up less by giving up our lives, if only because we will sacrifice fewer years of life.
6. There is less duty to die to the extent that you can make a good adjustment to your illness or handicapping condition, for a good adjustment means that smaller sacrifice will be required of loved ones and there is more compensating interaction for them. (However, we must recognize that some diseases—Alzheimer's or

Huntington's chorea—will eventually take their toll on your loved ones no matter how courageously, resolutely, or even cheerfully you manage to face that illness.)

7. There is more duty to die to the extent that the part of you that is loved will soon be gone or seriously compromised. There is also more duty to die when you are no longer capable of giving love. Part of the horror of Alzheimer's or Huntington's, again, is that it destroys the person we loved, leaving a stranger and eventually only a shell behind. By contrast, someone can be seriously debilitated and yet clearly still the person we love.

> In an old person, "I am not ready to die yet" does not excuse one from a duty to die... A duty to die seems very harsh, and sometimes it is, but if I really do care for my family, a duty to protect their lives will often be accompanied by a deep desire to do so...[18]

Remaining Perplexities and Practical Considerations

Assuming that the foregoing account is correct— that any of us might plausibly have a duty to die under fairly ordinary circumstances—some important questions still remain. Some of these are related in direct and practical ways to the recognition of a duty to die. Others bear less directly on the problem, but may affect our willingness to recognize the duty openly or in certain settings.

One clearly relevant question concerns the means of suicide available. If no legal, safe, effective, and emotionally acceptable means exist, beneficence calculations will rarely find active suicide to have the best consequences. Under such circumstances, a distinction between self-killing and allowing one's death would appear; we might have a duty to let ourselves die, but not to kill ourselves or have ourselves killed. On the other hand, if we suppose physician-assisted suicide to be available and socially acceptable, then active suicide will at times be the more beneficent choice.

If safe and convenient means of suicide become available, some further complications arise. One is the possibility that the wrong people will choose it or that it will be chosen for the wrong reasons. Undervalued and overconscientious people could be expected to recognize a duty to die too quickly, but others will, as they always have, continue taking too much and shirking their duties. Probably not much could be done about this. This problem does not distinguish a duty to die from any other duty and does not arise from its mere recognition. A more pressing problem concerns the likelihood that suicide choices would be manipulated in unacceptable ways.[19] If suicide joins the list of reasonable options for the old and ill, those who ought to be caring for them may be motivated to make the option more attractive by withholding full effort from other options.[20]

A broader problem arises here. Suppose that in the face of particular circumstances, a beneficence calculation shows that a person has a duty to die. Now suppose also that these particular circumstances are not as they ought to be; they arise out of an unjust social structure and corrupt, inefficient support systems in the community. What are we to think about a duty to die considered from this wider perspective? Surely, it would be within reason to say that the duty to die ought to be resisted in protest against the unfair conditions. It would make a difference how many people had to suffer, and in what ways, for this protest to be made. Clearly, though, the stronger duties may lie with those who have power to improve the conditions.

Finally, there are the familiar worries about slippery slopes. Some observers find the suggestion that there is a duty to die to lie at the bottom of a slope that had conscious health care allocation or a right to suicide at the top.[21] Others will worry that the duty as considered here will lie at the top of a slope, which has euthanasia programs to eliminate unwanted people with the duty to die strictly enforced, at the bottom. Though this worry would seem premature, it ought not to be forgotten entirely. As increasing numbers of people live longer and need more, we will not be able to avoid moral discomfort.

Notes and References

[1]Nat Hentoff (1997) Duty to die? *Washington Post,* May 31, v. 128, p. A1, col. 1.

[2]Richard Lamm (1984) Long time dying, *The New Republic*, August 27, pp. 20–23.

[3]John Hardwig (1997) Is there a duty to die? *Hastings Center Report* 27, no. 2, 34–42; Margaret Battin (1987) Age rationing and the just distribution of health care: is there a duty to die? *Ethics* 97:2 317–340.

[4]*See,* for example, Rosamond Rhodes (1998) Physicians, assisted suicide, and the right to live or die, in Battin, Rhodes, and Silvers, *Physician Assisted Suicide,* Routledge, New York, pp. 165–176.

[5]*See* Joel Feinberg (1970) The nature and value of rights. *J. Value Inquiry,* v. 4, 245–257.

[6]Liam B. Murphy (1993) The demands of beneficence. *Philosophy and Public Affairs,* pp. 267–292.

[7]Peter Unger (1996) *Living High and Letting Die: Our Illusion of Innocence,* Oxford, New York; F. M. Kamm (1993) *Morality, Mortality, v. 1: Death and Whom to Save from It,* Oxford.

[8]Leslie Pickering Francis (1998) Assisted suicide: Are the elderly a special case? in Battin, Rhodes, and Silvers, *Physician Assisted Suicide,* Routledge, New York, pp. 75–90.

[9]C. G. Prado (1998) *The Last Choice: Preemptive Suicide in Advanced Age,* 2nd ed., Praeger, New York.

[10]M. Pabst Battin and David J. Mayo, eds. (1980) *Suicide: The Philosophical Issues,* St. Martin's Press, New York.

[11]John Hardwig (1997) Is there a duty to die. *Hastings Center Report* 27, no. 2, 34–42.

[12]Merrill Matthews, Jr. (1998) Would physician-assisted suicide save the healthcare system money? in Battin, Rhodes, and Silvers, *Physician Assisted Suicide,* Routledge, New York, pp. 312–322.

[13]Anita Silvers (1998) Protecting the innocents from physician-assisted suicide, in Battin, Rhodes, and Silvers, *Physician Assisted Suicide,* Routledge, New York, pp. 133–148.

[14]Philippa Foot (1992) The problem of abortion and the doctrine of the

double effect, in John Fischer and Mark Ravizza, *Ethics: Problems and Principles,* Harcourt Brace Jovanovitch, Fort Worth, TX pp. 59–67.

[15]For example, *see* Raanan Gillon (1996) Intending or permitting death in order to conserve resources, pp. 199–207, and Norman Daniels (1996) On permitting death to conserve resources, pp. 208–215, both in Tom L. Beachamp, ed. *Intending Death: The Ethics of Suicide and Euthanasia,* Prentice Hall, Englewood Cliffs, NJ; *see also* Norman Daniels (1987) Just Health Care, Cambridge University Press, New York, 1985, and Daniel Callahan (1987) *Setting Limits: Medical Goals in an Aging Society,* Simon and Schuster, New York.

[16]Ronald Dworkin (1993) *Life's Dominion: An Argument About Abortion, Euthanasia, and Individual Freedom,* Knopf, New York.

[17]Thomas Hill (1973) Servility and self-respect. *The Monist* v. 57, no. 1, January, 87–104.

[18]John Hardwig (1997) Dying at the right time, in Hugh LaFollette, ed., *Ethics in Practice,* Blackwell, Cambridge, MA, pp. 53–65.

[19]Margaret Battin (1994) Manipulated suicide, in Margaret Battin, *The Least Worst Death,* Oxford, New York, pp. 195–204.

[20]David Mayo and Martin Gunderson (1993) Physician assisted death and hard choices. *J. Med. Philos.* 18, 329–341.

[21]Esther B. Fein (1996) The right to suicide, some worry, could evolve into a duty to die, *New York Times* sec 1, p. 24, col 1, April 7.

Abstract

In this paper, I argue that although we cannot defend a justification of a duty to die when it prescribes the sacrifice of one person for the greater good of others, we might defend a duty to die when the agents on whom it is imposed find it rational to consent to it. I try to show that it is rational for an agent to consent to a duty to die when he or she finds it in his or her prospective interest to adopt and comply with such a duty and when he or she likewise finds it in his or her prospective interest to accept the principles determining the situation in which he or she assesses the duty. Although a duty to die, in the same manner as other duties, might go against a agent's long- and short-term interests in the particular case, it must not go against his or her interests when viewed from the perspective of the cumulative consequences of general compliance with the duty over the course of his or her life.

The duty to die arises primarily in the context of costly medical care and military obedience. The duty is justified when general compliance with such a duty actually increases our prospects of survival overall and the circumstances in which we assess the duty are themselves consistent with principles that we accept. Therefore, we have a duty to forgo medical care when a general principle rationing that care actually increases our chances of survival over our lives and we accept the policies that lead to the shortage of care. Similarly, we have a duty to obey military commanders at the cost of our life when general compliance with a principle of obedience increases our prospects of survival overall and we accept the policies that lead to the war.

The Duty to Die: A Contractarian Approach

Robert E. Ehman

Two main approaches to justifying duties are the teleological and contractarian. The first justifies duties by the goals or goods to be achieved by compliance with the duty; the second justifies duties by consent to the rule imposing the duty on the part of the agent to whom it applies. To justify a duty to die in terms of the consequences, we need to show that it leads to more beneficial consequences overall than we can achieve without it; to justify a duty to die in terms of consent, we need to show that the principle imposing the duty is acceptable to the agent on whom it is imposed.

Although we shall find that a contractarian approach provides a more compelling justification for duties than a teleological approach, we shall also find that it is more difficult to justify a duty to die from this perspective than from the teleological perspective. For from the teleological perspective, to justify a duty to die, we need only establish that prospective consequences are beneficial overall, but from the contractarian perspective, we must show not only that an agent's prospective benefits from the rule prescribing such a duty exceed his or her prospective costs, but also that principles that determine the position from which he or she assesses these benefits are themselves acceptable to him or her for this same reason.

My paper will have three parts. The first part of the paper examines the meaning of a duty to die and distinguishes it from other rights and duties with which it might be confused. The second part explores the teleological justification of a duty to die and points up the limitations of that approach. The final section of the paper considers the conditions of a contractarian justification of a duty to die.

The Duty to Die

When we speak of a duty, we mean an action that we are obligated to undertake or to refrain from regardless of whether in that case we find it in our interest to do so. To the extent that we are permitted to do whatever we desire to do or find to be in our interest in a given case, we cannot be said to have a duty in that case. Duty is a constraint on the scope of permissible actions, but not all constraints are duties. When a gunman threatens my life in order to obtain my money, he imposes a constraint on me, but the constraint does not impose a duty. For in this case, the constraint is simply the requirement to take the means necessary for my survival. For a constraint to be a duty, it must be justified in terms other than our own self-interest in the particular case. For this reason, the imposition of external sanctions that make it in our interest to perform an action is not sufficient to make the action a duty, nor is the absence of such sanctions sufficient to free us from a duty to perform the action.

Moral duties are distinct from legal duties, and the necessary conditions of each are distinct. A necessary condition of the validity of legal duties is the validity of the procedure by which the law imposing them is enacted. Moral duties do not presuppose the validity of any legal procedure. An agent might find that a legal duty contradicts a moral duty, unless conformity with moral duty is among the criteria of the legitimacy of legal enactments.

A duty to die is a duty to perform an action that is likely to put an end to our lives or to refrain from an action necessary to save our lives. Although the question of a duty to die arises most often in cases of costly medical procedures, it also arises in war or in a situation in which we face the choice between saving our own lives or the lives others to whom we have special relationships.

When we raise the issue of a duty to die in a medical context, we are not normally speaking of a duty to take an action to end our lives, but rather a duty not to purchase, claim, or accept medical care necessary to prolong our lives. This must be distinguished from a right to die. The right to die imposes a duty on others not to interfere with our own decision to reject the medical care needed to preserve our lives or, in some cases, might impose a duty on the part of others to respect our decisions to end our life when the decision is an informed decision. The right to die is consistent with a right to claim the resources necessary to save our lives and, therefore, is independent of a duty to die.

The duty to die must also be distinguished from the moral or legal duty not to circumvent or attempt to violate the contractual terms of medical insurance to which we have agreed. If to save on insurance costs, we purchase insurance that denies benefits necessary to prolong our lives when we had the option of a more comprehensive plan and were informed of this limitation at the time of our choice, we cannot consistently demand that the insurance pay for the life-saving treatment that we have contractually agreed to forgo. The insurance company has no duty to provide these benefits, but we may still have a right to accept the benefits or to purchase them at our own expense. Our duty in this case is a duty to comply with our contractual obligations.

The duty to die must also be distinguished from a duty not to violate the rules of a public health plan in order to obtain life-saving treatment that is not provided under that plan. In Great Britain, the NHS fails to cover the cost of certain life-saving treatments for those over a certain age. These include dialysis and heart surgery for those over 55. The patients in these cases certainly have no legal right to the care and have a legal duty not to circumvent the rules of the plan. Since in the case of a public plan the patient does not contractually agree to the plan, the moral obligation to respect the rules of the plan is more problematic than in the case of a private plan to which he or she contractually agrees. However, even if he or she has a moral duty to comply with the rules of the public plan, he or she might still, as is the case in the United Kingdom, have a legal and moral right to purchase life-saving care with his or her own resources. Therefore, a moral duty to comply with the constraints of a public health plan does not by itself amount to a duty to die.

The duty to die must also be distinguished from a right to withhold certain life-saving treatment from those who are so incapacitated mentally or physically that they are no longer capable of informed consent and rational decision making. For in order to speak of an agent's having a duty, the agent must be capable of recognizing the duty, and able to choose to comply with it in the face of conflicting desires. This is not possible for those who are incompetent. The duties with respect to the treatment of incompetent patients are duties of the caregivers, not of the patients themselves.

In the same manner as a duty to die in a medical context, a duty to die in war is not typically a duty to end our lives, but one to act in such a manner as to put our lives at radical risk or a duty to refrain from doing something to save our lives. The paradigm case is a duty to obey a command that imposes a very high or even certain prospect of death. This must be distinguished from acts of heroism that surpass the requirements of duty.

The Teleological Justification of a Duty to Die

Teleological justifications take the form of demonstrating that the anticipated benefits to all of those affected by the action or by the rule of action in question outweigh the expected costs. There is dispute over whether we are to consider the consequences of each act, as act utilitarians propose, or to consider the consequences of a general compliance with a rule, as rule utilitarians maintain. However, both forms of teleological justification decide duties in terms of the overall consequences of the duty, not the consequences for each agent to whom it applies. Therefore, it permits one agent's interests to be sacrificed for those of others when the benefits to others outweigh the loss to the one.

The main objection to teleological approaches to moral justification is that these might justify duties that prescribe that one person sacrifice his or her interests simply for the benefit of others. For although teleological approaches take account of the prospective benefits and costs to each person affected by a duty, they might nevertheless justify duties that impose costs on a given agent that outweigh any expected benefits, not only in the particular case, but over the course of his or her life. Unless this person has an overriding concern for the welfare of those benefited, it is hard to explain why he or she would find it rational to accept a duty that benefited others at his or her expense. Therefore, although apart from such a concern for others the agent might be coerced into sacrificing himself or herself for others, he or she cannot have a duty to do so, for a duty must be able to motivate apart from an agent's interests in the given case. However, coercion in contrast to a duty motivates only by appeal to an agent's interests in the case in which it applies.

John Harsanyi argues that once an agent assesses principles from an impartial perspective of radical uncertainty regarding how the principles will affect him in fact, he will find it rational to accept a utilitarian principle even though that might impose duties that promise to disadvantage him in fact over the course of his life.[1] For Harsanyi, it is appropriate to assess moral rules from this perspective, since it removes bias from the choice of moral rules. From this perspective, an agent will find it rational to assume that he has an equal probability of being in any position in the society, since there is no reason to suppose that any position is more probable than another. If he assumes equiprobability, he will for this reason regard his prospects as the average prospects of the society. Since a principle maximizing these average prospects will then maximize his own prospects, he will find it utility-maximizing to accept a principle that maximizes average prospects. This is a principle of maximizing average social utility.

However, as John Rawls has pointed out, even if we accept the assumption that principles must be assessed from an impartial standpoint of radical ignorance of our particular situation, it is not rational to assume equiprobability of gain. For Rawls, it is not rational to assume this because it amounts to a gamble with regard to fundamental principles without any information concerning the odds of their effect on us. From a position of radical uncertainty, it is instead rational according to Rawls to adopt a maximin principle of choice that maximizes benefits in the worst case regardless of the probability of that case. For this ensures against the worst eventualities when we are unable to calculate their risks.

Rawls' proposal that we adopt principles that maximize the worst case might itself appear to be a teleological justification of moral duties. For although maximizing the worst case is not a utilitarian principle, it appears to prescribe an end result. However, the maximin is not a general goal to be achieved, but rather a constraint on the achievement of any overall goal. For Rawls, the principles of duty are not means to the maximization of worst cases, but rather constraints on actions that maximize social goods at the expense of those worst off. The maximin is for this reason not so much a social goal as a constraint on the pursuit of social goals.

There is indeed, as Rawls implies, a serious question whether it is even possible in good faith to commit to a principle imposing a duty to die simply for the benefit of others in the light of the strains of commitment that it would impose on us. Rawls argues that we cannot rationally commit ourselves to a loss of freedom or a less than equal opportunity for education and culture for the greater welfare of society, since the strain of complying with that commitment would be too great. The argument from the strain of commitment is even more convincing when applied to the commitment to give up our lives for the good of others prescribed by a teleological justification of a duty to die.

The Contractarian Justification of a Duty to Die

The point of contractarian justifications of duty is to address the problems that arise from justifying moral principles by appeal to the aggregate benefits of complying with them. The contractarian attempts to justify duties by establishing that it is in the rational interest of each agent to accept the duties, not simply in the aggregate interest of all of those affected to do so. The contractarian denies the legitimacy of duties that require the sacrifice of one person simply for the good of others.

This raises doubt about whether there can be a contractarian justification of a duty to die. For unless the beneficial consequences for others of the imposition of a duty to die are sufficient to make that duty acceptable, there might appear to be no justification of a duty to die. A duty to die appears to make one's life disposable for the good of others and to deny the autonomy necessary to distinguish duty from what we are simply compelled to do.

A duty, as we have already pointed out, might prescribe that we act independently of our interests in particular cases. These interests might include an interest in the prolongation of our life. The question of whether a principle prescribing a duty to die is rationally acceptable is essentially the same as that of whether principles prescribing other duties are rationally acceptable. The question is how it can ever be rational for an agent to obey a principle that imposes a cost that does not promise in that case a benefit worth that cost.

In many cases, we find it in our interest to comply with a principle that benefits us cumulatively over the course of our lives, even though it prescribes actions that are not individually cost-beneficial. For an example, we may take a rule against smoking. This might in each case impose a cost on the smoker that exceeds the benefits of the act. The long-term benefits of forgoing each cigaret might be lower than the costs in present distress. However, it may nevertheless be in the interest of a smoker to comply with such a rule. For although he might well lose more than he gains from forgoing each cigaret considered by itself, his commitment to comply with the general prohibition against smoking may gain him more than he loses. The benefits arise from general obedience to the rule, not from the consequences of each act taken by itself. The act itself might impose a net cost on the agent unless he or she complies with a rule against future smoking.

This same sort of argument can be used to justify moral principles. For although in particular cases, we might lose more than we gain even over the long term from a particular act of truth-telling or meeting of appointments, we might lose more than we gain from a practice of lying or breaking appointments whenever it appears to be to our advantage to do so. For example, a practice of truth-telling might prevent an otherwise disastrous destruction of our credibility whose risk we cannot accurately assess when we consider each act of lying in and by itself apart from other acts of lying in similar situations. There might be little risk of loss from each act of lying considered by itself. However, the cumulative effect of the practice of lying whenever it is in our interest to lie might lead a risk to our credibility larger than that of any of the particular acts. For an example of this mode of reasoning, we might take the practice of insurance companies that charge rates depending on the amount of driving as well as the risk profile of the driver. The insurance industry recognizes that the more one drives, the higher one's overall risk of accidents, even though the greater amount of driving does not increase the risk of any particular trip.

This same sort of reasoning might apply to military obedience or health care rationing. If we obey commanders only when we expect it to increase our prospects of survival in particular cases, we might so undermine military discipline that we decrease our prospects of surviving. Similarly, if we receive life-saving care whenever it promised to be beneficial in the particular case, we might so deplete the resources available for health care that we would diminish our overall prospects for care. Therefore, we might find a practice of obeying our officers or forgoing certain life-saving medical care in our interest as a practice, even though in particular cases, it is not in our interest.

There is an important distinction between principles that prescribe our actions toward ourselves, as do rules against smoking or eating certain foods, and moral rules that prescribe how we interact with others. In order for rules that govern our own behavior to be beneficial, we must make a practice of complying with them. However, there is no need for others to do so. On the other hand, in order to gain from rules that govern our interactions with others, both ourselves and others must make a practice of complying with them. For if others lie to us or fail to make appointments with us whenever it appears to their advantage, this diminishes the benefit that we obtain from our own compliance with the rule and might outweigh it. Therefore, apart from general compliance with rules of obedience or health care rationing, we might not benefit from complying with them and therefore be relieved of the duty to do so.

In order for an argument from the prospective benefits of the general acceptance of a duty to justify that duty, it is necessary to take a step back and ask why an agent is faced with the choice between a principle that might require him or her to act against his or her interests in a given case or else face an even higher risk of loss over his or her life. The choice of principles of duty often presupposes a situation that is itself affected by choices. If an agent does not accept the principles that give rise to the situation in which he or she finds it in his or her prospective interest to accept a given principle, that principle is no more consensual than a principle of complying with all demands of gunmen. For although compliance with a principle of complying with gunmen might be beneficial, it is not consensual, since it presupposes a situation in which others act in a manner that violates principles that we find it rational to accept.

The possible case of a duty to risk our life to save a person for whom we have a responsibility from criminal threats might appear to present a counterexample to the claim that we have no duties when these duties are rational only because of actions of others who violate principles we can accept. For we might have a duty to risk our lives to save a beloved person from hostage-takers, even though we reject the right of the hostage-takers to hold the person hostage. Therefore, although we do not accept the principle that leads to the situation, we may still have a duty to save the victim. However, the principle of risking our lives to protect our loved ones might not in this case be based on a situation that violates our principles. The principle might be one that we freely assume toward the person independent of bribes and threats, even though it is a threat to the person that makes it applicable to the given situation. This is to be distinguished from a principle prescribing that we comply with demands of gunmen, for that is a principle forced on us by a threat that we reject the right of the gunman to make.

Soldiers who reject the principles of the decisions that lead to a war might find it rational to obey their commander, but they do not have a duty to obey and to die for their country. Similarly, patients who reject the allocational principles that lead to the shortage of care that makes it in their interest to accept a principle that denies them certain life-saving care in order to provide an acceptable level of health care throughout their lives might find it rational to accept the limitation on care, but they have no duty to accept that constraint.

The amount of health care available in a society depends both on decisions with regard to the proportion of the resources of a society to be allocated to health care and on decisions that affect the level of resources that the society possesses. The more technologically advanced and wealthier the society, the greater are the resources of the society. The greater the resources overall, the greater are the resources available for allocation to health care. The greater the resources actually allocated to health care, the less likely it is that an agent will face a choice between a limitation of life-saving care and an unacceptable level of care in general. If an agent does not agree with the decisions that determine the level of health care in the given society, he or she cannot be said to consent to the conditions to which they lead. Therefore, he or she will find constraints that depend on these conditions to be imposed on him or her against his or her will.

In appealing to the acceptability of a principle, we are not simply asking whether agents in fact accept it. We are asking whether it is rational for them to do so. This raises more issues than we can fully resolve here, but in order to defend a contractarian justification of duty, it is necessary to give some account of rational acceptance. For economics and game theory, to be rational is to maximize our utility in the sense of a measure of our considered and informed preferences, but this maximizing principle fails to recognize that diverse, equally informed agents might have diverse risk preferences and diverse time preferences. Agents who are risk-averse, willing to gamble against the odds, or who prefer nearer or more distant phases of their lives might not be mistaken or ignorant with respect to these choices. For this reason, these choices are not as such irrational. To be rational, agents need not be risk-neutral or have an absence of preference for diverse phases of their lives, but simply to act on the basis of the relevant facts with regard to their choices. Moreover, they need not know everything relevant to their choices; they need only know what they can learn at a cost that they find worth paying from the perspective of what they already know.

In making rational acceptance of the principles that determine the situation in which we find ourselves a condition of genuine duty, we appear to make an irrational demand to base our duties on a hypothetical situation rather than on the situation in which we in fact find ourselves. We may face a war or a shortage of health care whether we agree with the decisions that lead to these or not. What is the point of making a distinction between what it is rational to do in that situation and what it is our duty to do? We often have no control of the decisions that determine our situation.

However, if we justify principles by reference simply to what is rational to accept in the actual status quo, we will not be able to distinguish between a principle that we are forced to accept and an autonomous principle that we find it rational to adopt apart from bribes or threats. The principle of complying with the demands of gunmen is not a principle of duty, because we would not find it rational to adopt and comply with it except for threats that we do not accept the right of the gunman to make. For a principle to be autonomous, it need not, as Kant supposed, be totally independent of an agent's interest since in this case, it is hard to see how an agent could make rational decisions with regard to principles at all. However, it must not be rational to accept only because of actions that violate principles agents find it rational to accept.

The decisions that lead to war or to the level of health care available in a society are typically collective decisions. These are not unanimous as are decisions in an ideally competitive market. To the extent that collective decisions affect the conditions under which it is rational to accept a duty, the acceptance of that duty depends on factors that are not altogether under the agent's control and with regard to which agents might disagree with each other. This has the consequence that different agents under the same conditions might not have the same obligations to conform to the results of the collective decisions. Those who endorse given military and foreign policy decisions or endorse given economic and health care policy decisions must rationally accept a duty to die when those decisions lead to a situation in which it is in their interest to accept a principle imposing that duty. On the other hand, those who reject those decisions might not find it rational to accept that duty.

In constitutional democracies, collective decisions are made in accordance with constitutional rules that permit each agent to have a vote and a voice. If agents find it rational to accept these rules for decision making, they will for this reason find it rational to accept the results of these procedures, even though they would not accept the results in and by themselves apart from the consensual constitutional procedures by which they are obtained. They find it rational to accept constitutional rules that permit nonunanimous collective decision making when it in their interest to permit nonviolent resolution of policy disputes in the absence of unanimity.

The appeal to consensual procedures answers socialist critics of markets who argue that unless market participants consent to the property rules of the market, market transactions are themselves coercive and violate the genuine moral rights of participants. These critics are correct that the rules of property that establish the parameters of market transactions may not be in themselves consensual. However, they are mistaken to reject the moral duty to respect property rights when the laws that govern the markets arise from constitutional procedures that the agents do accept to make collective decisions.

The acceptance of constitutional collective decision making procedures also answers libertarian critics of redistributive taxes and public goods decisions that impose costs on agents to which they do not consent. Both the socialist and libertarian fail to recognize that when we consent to the constitutional procedures that determine property rights, we consent to the property rights that result from them even though we would not consent to them apart from the consensual constitutional procedures. The argument that property rights or taxes are coercive and nonconsensual, therefore, applies to states that decide on these independent of the consent of the governed, but not to those who decide them on consensual principles.

However, consent to the results of constitutional procedures becomes problematic when the decisions impose a high risk of death. In recognition of the problematic character of legal requirements that impose a high risk of death, common law generally permits violations of otherwise legitimate property rights when necessary to save our lives. The rational point of accepting procedural resolutions of conflicts is, as Thomas Hobbes pointed out, to protect and preserve our lives. For this reason, when the outcomes of the constitutional procedures threaten our lives, these become coercive unless we consent in their own right to the specific rules and principles that put our lives in jeopardy. Although we need not agree in this manner to each of the principles that determine markets and public goods when our lives are not in question, we must where they are in question. The duty to die cannot be imposed on those who do not unanimously agree to it.

However, even when we do not find it rational to accept the decisions that lead to a war or shortage of health care, we might benefit from general compliance with principles that demand obedience to commanders or restriction of costly life-saving care. If we have benefited from these principles, are we not obligated to respect them? For if we are not, we are permitted to be a free rider who obtains benefits from others without paying our share of the cost. In markets, we are not required to pay for benefits when we do not purchase them and thereby accept the terms under which they are provided. In the case of public goods, we are required to pay for benefits that we might not purchase at that price as long as we agree to the decision-making procedures. However, as we have seen, the consent to decision-making procedures does not extend to cases where these take our lives. For this reason, we have no more obligation to pay for benefits we receive as a result of these than we have to pay for market goods that are provided against our will or pay for public goods that are decided by means of decision-making procedures we reject.

Conclusion

In assessing public policy decisions that affect in certain cases whether we live or die, we must consider whether we are willing to accept the results of the decision when they might lead to a situation in which it is rational to accept a principle that requires our death. In foreign policy, we must ask whether we are willing to pursue a policy at the risk of a war in which we might be called on to die; in economic and health care policy, we must ask whether we are willing to pursue policies that lead to a situation in which we might be called on to forgo certain life-saving treatment in order to provide a satisfactory level of care for ourselves and those we care about throughout the course of our lives and their lives. Those who reject the decisions that lead to these situations are free from a duty to die in these situations, even though they might find it prudential to comply with principles that require them to act in a way that leads to their death. There are only two conditions under which agents have a moral duty to die. The first is when it is both in their prospective interest to comply with a principle that might call on them to die in a given situation and in their prospective interest to support the decisions that lead to that situation; the second is when no decisions could have saved them from a situation in which it is in their prospective interest to comply with a principle that might impose on them a duty to die.

Notes

[1]For Harsanyi's position, *see* J. C. Harsanyi, (1953) Cardinal utility in welfare economics and in the theory of risk taking, *J. Political Economy*, 61. For Rawls' criticism of Harsanyi, *see* J. Rawls, *A Theory of Justice* Harvard University Press, Cambridge, pp. 167–183.

Abstract

I consider two central arguments against extending the right to die to include a right to active euthanasia. What have come to be called "wedge arguments" appeal to the "unstable" distinction between human life that may be actively terminated and human life that may not be actively terminated. I locate two possible sources of the instability: (1) the inherent vagueness of moral proscriptions against active euthanasia, and (2) the epistemological limitations unique to applying principles proscribing active euthanasia. Each version of the wedge argument is circular in ways that are not so subtle. I consider next a rule utilitarian argument that appeals to the corrosive consequences of any social policy liberalizing the right to die. I show that even if the effects are as bad as anticipated, the rule utilitarian argument fails. The argument makes the subtle but common rule utilitarian mistake of assessing social policies by appeal to our current attitudes, desires, and dispositions. What matters to the assessment of social policies of any sort are rather the attitudes, desires, and dispositions we would have were such a policy instituted. I conclude that wedge arguments and rule utilitarian arguments of this sort are seriously flawed, and cannot be convincingly arrayed against passive euthanasia, active euthanasia, or any reasonably restrictive duty to die. The very same internal flaws will resurface in each context of application. The most reasonable response is a cautious rejection of the arguments against the liberalization of social policies in these areas.

Rule Utilitarianism and the Right to Die

Michael Almeida

Competent adults who express a preference not to have further treatment administered should their conditions deteriorate to a persistent vegetative state are paradigmatic among those whose choices seem to enjoin the assistance of physicians. Among those willing to acknowledge a right to die, we should expect most to converge on the conservative belief that the right to die is a right of competent adults exercising their autonomous choice in dire circumstances. But of course, incompetent adults in equally dire circumstances might have no advanced directives and might never have articulated a (considered) preference regarding the cessation of medical treatment. We should expect fewer people to converge on the view that interested third-parties (whose informed judgment may consist of an impression of the patient's desire) may exercise the right to die on their behalf. Still fewer will agree that persons or "near-persons" that compose the never-competent, profoundly handicapped infants, for instance, implicitly defer judgment on the value of their lives to disinterested professionals.

The strict evidentiary restrictions on permissibly ceasing treatment in the absence of direct and explicit authorization reflects a broad suspicion that liberalizing the right to die would adversely affect our attitudes toward some important moral prohibitions. It has been noted, for instance, that unconstrained representatives will find themselves unable to resist authorizing the termination of a patient's life for reasons unrelated to his or her welfare.[1] More sinister still, we are cautioned that the right to die will be cynically asserted by bright and healthy people on behalf of unproductive and incurable patients whose moribund condition they find repulsive.[2] It is anticipated in general that liberalizing the right to die would gradually erode our sensibilities about the value of human life and result in a continual and inexorable diminution of human well-being.

The reasons adduced above are most forcefully arrayed against extending the right to die to include the right of competent adults to active euthanasia. Indeed the anticipated consequences of this liberalization are so grave that there exist no evidentiary standards, strict or otherwise, whose achievement would make permissible actively killing a patient. A fully competent adult ideally informed of his or her current and impending condition is morally and legally prohibited from exercising his or her free will in the authorization of active euthanasia.

There is no convincing argument for the moral injunction prohibiting active euthanasia based on the intrinsic disvalue of actively killing. Small or no question remains that the specific act of killing a patient is in many instances no worse than letting a patient die, provided all involved are well-motivated, competent, and ideally informed. Also, apart from this, the very metaphysical distinction between actively killing and letting die remains a vexed point. Metaphysics aside, killing might well be just what the law says it is.

The most interesting arguments for the moral injunction against active euthanasia concede that in many circumstances, specific acts of killing do not differ in moral status from specific omissions resulting in death. The basis of the moral injunction is rather the deleterious consequences of a social policy making permissible actively taking human life. It is expected that even highly qualified social policies permitting active euthanasia would have corrosive consequences far worse than those accompanying the current social policies permitting the terminal cessation of treatment.

I consider in Worries About Wedge Arguments what is now entitled the "wedge argument" against the institution of a social policy extending the right to die to include a right to active euthanasia. According to the wedge argument, the distinction between human life that may be actively terminated and human life that may not be actively terminated is "unstable." Principles permitting the active termination of life will become increasingly inclusive, and the class of those individuals who may be actively killed will gradually expand.[3] The legitimacy of such fears is confirmed, we are told, in recent historical record. The darker regions of the Nazi era, for instance, illustrate how an initially well-intentioned program of euthanasia can deteriorate precipitously into a method of terminating political enemies and scapegoats.[4]

Despite the strong rhetorical force of the wedge argument, I show in Worries About Wedge Arguments that the argument fails. The argument assumes either that moral proscriptions specifically against active euthanasia are unavoidably vague, or that the epistemic limitations of moral agents make impossible a precise application of any moral proscription against active euthanasia. I show here that there are good empirical reasons contravening each assumption. Indeed we will find that there is no noncircular basis for either of these assumptions.

In The Rule Utilitarian Argument, I examine a more ambitious rule utilitarian argument against extending the right to die to include a right to active euthanasia. The rule utilitarian argument appeals to the effects of social policies on our desires and dispositions. A social policy permitting even restricted forms of active euthanasia would inure moral agents to the general practice of active killing. The argument contends that the respect for policies protecting human life will decrease, and less restricted policies on active euthanasia will be implemented.

I show in here that the expected consequences of a social policy permitting active euthanasia provide no utilitarian reasons against instituting the policy. Even if the consequences are just as anticipated, it is incorrect to conclude that there is a rule utilitarian prohibition against enacting a policy permitting active euthanasia. The argument makes the subtle, but common rule utilitarian mistake of assessing social policies according to our current desires, dispositions, and attitudes. In fact, our current desires, dispositions, and attitudes are not relevant to a rule utilitarian assessment of a policy permitting active euthanasia, or any other social policy. Even if the consequences are exactly as anticipated, they might well maximize overall utility, contrary to the rule utilitarian conclusion.

I conclude that the arguments advanced against liberalizing the right to die depend on nothing more than rhetorical force. The prospect of profound and dismal consequences can move us to reject social policies our better judgment would have us accept. I conclude that even if we grant the tendentious point concerning the dire consequences, the rule utilitarian argument still fails, and the wedge arguments fare no better. However, notice finally that the arguments arrayed against the right to die are generally assumed to apply *a fortiori* against the duty to die. Perfectly analogous flaws reappear in the contexts of passive euthanasia, active euthanasia, and in any reasonably restrictive duty to die. These arguments fail to undermine social policies making permissible each of these practices.

The most reasonable response to the arguments against a social policy permitting active euthanasia or against acknowledging a duty to die is a cautious rejection. Perhaps there are better versions of the arguments. Perhaps there are just better arguments. But as things stand, there appears to be no reason against instituting the more liberal policies.

Worries About Wedge Arguments

The initial difficulty in evaluating wedge arguments is that their structure is generally not well articulated. Consider, for instance, the following version of the wedge argument.

> [A]lthough active euthanasia might be morally permissible in cases in which it is unequivocally voluntary and the patient finds his or her condition unbearable, a legal policy permitting euthanasia would inevitably lead to active euthanasia being performed in many cases in which it would be morally wrong. To prevent those other wrongful cases of euthanasia, we should not permit even morally justified performances of it.[5]

Elsewhere we find the following similar version of the argument:

> Uncertainty and mistrust are already too much a part of the healing relationship. Euthanasia magnifies these ordinary and natural anxieties. How will the patient know whether the physician is trying to heal or relinquishing the effort to cure or contain illness because she favors euthanasia, devalues the quality of the patient's life or wants to conserve society's resources? The physician can easily divert attention from a "good" death by subtly leading the patient to believe that euthanasia is the *only* good or gentle death...We cannot simply say the "good" doctor would never abuse the privilege of euthanasia. Whose agent is the doctor when treatment becomes marginal and costs escalate? Will the physician's notion of benevolence to society become malevolent for the older patient?...Can patients trust physicians when physicians arrogate to themselves the role of rationers of society's resources or are made to assume that role by societal policy? We already hear much talk of the social burdens imposed by chronically ill, handicapped, terminal adults and children and the necessity of rationing with physicians as gatekeeper.[6]

Each of the arguments appeals to the probable abuses of any social policy granting patients a right to active euthanasia. The anticipated abuses are no doubt alarming, but precisely why such policies would "inevitably lead" to widespread abuses is obscure. In spite of the unsettling tone of these arguments, it is unreasonable and worse to base wedge arguments on some general moral indictment of the medical profession. Certainly any plausible version of the argument must assume generally well-intentioned and well-informed patients and physicians. However, how could well-informed and well-motivated physicians misapply highly qualified moral (or legal) principles permitting active euthanasia?

It has been advanced in favor of wedge arguments that principles permitting active euthanasia would lead to widespread abuse not because of ill-motivated physicians, but because of the inherent vagueness of prohibitions against killing. Therefore, for instance, we find the following more interesting possibility.

> Proponents of wedge arguments believe that the initial wedge places us on a slippery slope for at least one of two reasons... It is said that our justifying principles leave us with no principled way to avoid the slide into saying that all sorts of killings would be justified under similar conditions. Here it is thought that once killing is allowed, a firm line between justified and unjustified killings cannot be securely drawn. It is thought best not to redraw the line in the first place, for redrawing it will inevitably lead to a downhill slide.[7]

Conceding for the moment that principles prohibiting active euthanasia are inherently vague, there seems little question that the foregoing description overstates the difficulties. The problem with initiating a social policy permitting euthanasia is not that we will be left unable to draw a firm line between justified killing and unjustified killing. We cannot, for instance, reasonably anticipate any uncertainty about the morality of killing an otherwise healthy adult seeking medical assistance for a broken thumb. The difficulty must fall in a comparatively smaller range of cases.

On the other hand, the difficulty cannot be with the morality of killing in situations relevantly similar to those in which active euthanasia is paradigmatically permissible. If we are considering situations that are in fact relevantly similar, then there might be psychological reservations about the permissibility of active euthanasia, but it is difficult to see how there could be any logical reservations. Moral intuitions aside, consistency alone would dictate treating similar cases similarly.

However, the wedge argument might have been designed to exhibit a more complex logical structure. Let us assume that there is a paradigm situation S_1 in which active euthanasia is morally permissible. Let $S_1, S_2, \ldots, S_n$ be a sequence in which each situation S_j is pairwise relevantly similar to its successor, S_{j+1}. Therefore, S_1 is relevantly similar to S_2, S_2 is relevantly similar to S_3, and so on. Now, since active euthanasia is morally permissible in S_1 (by hypothesis), and S_1 is relevantly similar to S_2, we know that active euthanasia is morally permissible in S_2. Repeated application of this reasoning of course commits us to the view that active euthanasia is morally permissible in S_n. Now, of course, S_n is relevantly similar to S_{n-1}, but it might be contended that S_n is not relevantly similar to S_1.[8] The problem with initiating a policy of active euthanasia, then, is that it would result in permissible killing in situations relevantly dissimilar to the paradigm case.

Assuming the wedge argument is a slippery slope of this type, any convincing defense of the argument would require, at least, a plausible instantiation of the schema described above. The typical presentation of the wedge argument does not even approximate the precision necessary for a cogent slippery slope argument. Broad characterizations of historical precedent, such as the deteriorating events in the Nazi era, certainly have rhetorical force, but they are obviously poor instantiations of the schema above. In the absence of a detailed description of some reasonably anticipated slippery slope, the argument remains unconvincing.

However, notice that there are independent reasons to believe that a convincing slippery slope argument against instituting a policy permitting active euthanasia is not forthcoming. It is central to the argument that the relation "relevantly similar to" be intransitive, but there is nothing in wedge arguments that begins to demonstrate this. Also, even if we suppose that the relation were intransitive, it remains unlikely that a convincing slippery slope argument is forthcoming.

Suppose that "relevantly similar to" were an intransitive relation. Now, what explains the fact that there have been no slippery slope problems in the application of principles permitting passive euthanasia? If the relation were intransitive, then of course it would not matter whether we were deliberating about the proper application of passive euthanasia or the proper application of active euthanasia. Moral deliberation ought to have resulted in the conclusion that passive euthanasia is permissible in situations relevantly dissimilar to some paradigm situation, and so ought to be generally prohibited. However, there is nothing to confirm such fears about passive euthanasia. Therefore, it is reasonable to conclude that a convincing slippery slope argument against active euthanasia is unlikely.

However, consider an alternative argument appealing to the epistemic limitations of moral agents. Epistemic limitations could of course prevent a precise application of any policy permitting active euthanasia, however precisely stated the policy. Even if there is a morally relevant difference between each situation S_j and its successor S_{j+1}, we might not be in an epistemological position to perceive the difference, and this could be the source of another type of slippery slope argument.

The problems with the epistemic account of the slippery slope argument are perfectly analogous to those noted above. It simply begs the question to assume that moral agents encounter epistemic limitations in the application of principles or policies permitting *active* euthanasia, but encounter no such difficulties in the application of principles or policies permitting passive euthanasia. Also, again, there is little historical evidence to confirm any worries about an epistemological slippery slope in the application of principles permitting passive euthanasia.

The proposal that we extend the right to die to include the right to active euthanasia appears to arouse fears of potential abuse that find expression in various forms of the "wedge argument." However, the metaphysical version of the wedge argument, so far forth, lacks the rigor and detail necessary for a convincing slippery slope argument. And, it appears unlikely that a rigorous and detailed version of the argument is forthcoming. The epistemic versions of the wedge argument immediately encounter problems of circularity. And the circularity is not particularly subtle. There is simply no reason to assume that there are unique epistemological limitations in deliberating about active euthanasia. As a result, it appears that no version of the wedge argument supplies compelling reason against extending the right to die to include a right to active euthanasia.

The Rule Utilitarian Argument

Rule utilitarians in general maintain that the proper application of the principle of utility is somewhere above the level of particular actions. However, there are otherwise important differences among the rule utilitarians about the precise application of the principle. The principle of utility is sometimes applied exclusively to general practices, but sometimes it is applied to individual rules, and sometimes to complete moral codes.[9] However, rule utilitarians also differ over whether it is the adherence to a rule, the acceptance of a rule, or merely the institution of a rule whose utility is supposed to be evaluated. All of these distinctions can make a substantial difference in the deliberations of rule utilitarians. Even if the general adherence to a particular rule would have untoward consequences, for instance, the consequences of merely instituting the rule might well be harmless.

Now consider a complete moral code that includes a policy permitting a highly qualified form of active euthanasia. Assuming that the moral code is otherwise acceptable, our general adherence would yield undesirable consequences only if physicians and the general public preferred a more liberal policy on the right to die.[10] At best, then, the conclusion of this rule utilitarian argument is that we need not liberalize the right to die beyond a generally preferred, but highly restricted, form of active euthanasia.

However, consider an alternative rule utilitarian argument. Suppose the principle of utility is applied to the institution of specific rules rather than to the adherence to complete moral codes. The moral concern is now restricted to the overall utility of instituting specific rules. Note that no assumptions are made concerning whether the institution of a social policy will result in either general acceptance or general adherence to that rule, or for that matter any other rule constituting the complete moral code. What are the anticipated consequences of instituting a social policy extending the right to die to include the right to active euthanasia?

Among the reasonably anticipated consequences, some certainly appear to be favorable.[11] We can expect, for instance, that some profound and otherwise uncontrollable suffering will be eliminated. We can also expect the utility forthcoming respecting the autonomy of individuals desiring to die in a way commensurate with their dignity. But rule utilitarians have argued that we can also reasonably anticipate consequences of considerable disutility.

Introducing a form of legitimate killing into one's moral code would have a deleterious effect on our attitudes toward other moral rules demanding respect for human life. Some utilitarians maintain that instituting a policy of active euthanasia would, for instance, alter our attitudes toward involuntary active euthanasia. They maintain that we might reasonably anticipate that the destruction of defective newborns will become a *common* and *accepted* practice.[12] We can also expect that as the population increases, there will be more and more neglect for the aged. To enumerate still other consequences, we will find that capital punishment will be more liberally applied, that physicians will find it more tempting to use lethal injections in questionable circumstance, that the general prohibition against killing noncombatants will be even less respected that it is currently, and so on.

According to this rule utilitarian argument, the prospect of instituting a policy liberalizing the right to die appears dismal, and the conclusion that has been drawn is that there are good utilitarian reasons against instituting such a liberal policy. However, the conclusion is mistaken. I note first that there are important worries about the actual psychological effects (the anticipated effects on our moral attitudes) of instituting such a social policy. The effects might be less severe than anticipated. Moreover, there are additional worries about what we would actually do, were we to develop such moral attitudes. But quite apart from the inevitable disputes about the foreseeable future, there are serious problems facing this rule utilitarian argument.

Let us suppose that the institution of a liberal policy on the right to die would actually have the anticipated "deleterious effects" on our moral attitudes. Let us also grant that our "relaxed attitudes" toward rules demanding respect for human life would have us engaging in just the sorts of practices described above.[13] It is simply a mistake to conclude from this that instituting a rule liberalizing the right to die would fail to maximize overall utility.

Give our actual attitudes, desires, and dispositions, it is true that any social policy lessening our respect for human life appears abhorrent, but the overall utility of instituting a social policy liberalizing the right to die is not properly assessed by appeal to our current attitudes, desires, and dispositions. The rule utilitarian assessment of any social policy must appeal to the attitudes, desires, and dispositions we would have were such a social policy instituted. According to the rule utilitarian argument, the anticipated behavior is a consequence of a presumably widespread change in the attitudes of moral agents. Even if we suppose that the change in attitudes, desires, and dispositions is at profound variance with current attitudes, desires, and dispositions, rule utilitarians cannot evaluate those changes *a priori*.

The conclusion of the rule utilitarian argument is not that instituting a policy liberalizing the right to die would have dismal consequences. The conclusion is rather that instituting a policy liberalizing the right to die would alter, perhaps dramatically alter, the kinds of practices maximizing overall utility. Reasoning as rule utilitarians, it is not to the point that we dislike, or even deeply dislike, the attitudes we will develop as a result of instituting the social policy. The dislike might well be reciprocated. Reasoning as utilitarians, the question is whether those anticipated changes in attitude are consistent with the maximization of overall utility. Also, as the rule utilitarian argument is presented, there is no reason to think that instituting a policy liberalizing the right to die will fail to maximize overall utility.

Conclusions

The arguments advanced against liberalizing the right to die to include the right to active euthanasia appear to depend on nothing more than rhetorical force. There is, of course, considerable force in appealing to the dramatic and dismal consequences of instituting a social policy permitting active euthanasia. However, it is not obvious that anything nearly so severe would result from instituting such a policy. And if we grant the tendentious point concerning the dire consequences, the rule utilitarian argument still fails. The wedge arguments fare no better. There is no reason to anticipate, moreover, that applying these arguments against a policy acknowledging a duty to die would have any more success. The very same flaws reappear in contexts of passive euthanasia, active euthanasia, and in any reasonably restrictive duty to die.

A cautious rejection seems the most reasonable response to the arguments against liberalizing these social policies. Perhaps there are better versions of the arguments. Perhaps there are just better arguments. But, as things stand, there appears to be no reason against instituting a more liberal set of policies.

Notes and References

[1]*See* Cruzan v. Director (1990) Missouri Department of Health, *United States Supreme Court Reports* 497, 261–357.

[2]Cf. Leon Cass (1993) Is there a right to die? *Hastings Center Report* 23, pp. 34–43.

[3]Tom Beauchamp (1978) A reply to Rachels on active and passive euthanasia, in Tom Beauchamp and Seymore Perlin, eds. *Ethical Issues in Death and Dying.* Prentice-Hall, Englewood Cliffs, NJ, pp. 246–258.

[4]*See* Tom Beauchamp, A reply to Rachels on active and passive euthanasia, ibid., esp. sec. III.

[5]*See* Dan W. Brock (1992) Voluntary active euthanasia, *Hastings Center Report* 22, No. 2, 10–22. Brock here reiterates a version of the wedge argument, but does not fully endorse the argument. I elaborate below a version of the argument Brock does find compelling.

[6]*See* Edmund D. Pelligrino (1994) Euthanasia as a distortion of the healing relationship, in Tom Beauchamp and Leroy Waters, eds. *Contemporary Issues in Bioethics.* Wadsworth, CA, pp. 483, 484.

[7]Cf. Tom Beauchamp, A reply to Rachels on active and passive euthanasia, op. cit. We find here Beauchamp alluding to wedge arguments based on an appeal to vague principles.

[8]The argument assumes of course that "relevantly similar to" is a symmetrical, but intransitive relation. I discuss this below.

[9]Compare, for instance, the complete-moral-code version of rule utili-

tarianism endorsed by R. B. Brandt (1992) *Morality, Utilitarianism, and Rights.* Cambridge University Press, Cambridge.

[10]Note the earlier assumption that the concern about the extending the right to die to include a right to active euthanasia is not that there are no circumstances under which active euthanasia is intuitively permissible. The objection is rather to the consequences of a social policy allowing even intuitively permissible forms of active euthanasia. If we are assessing the consequences of general adherence to the highly qualified principle, the social policy objection is immediately undermined: we simply assume that no one will fail to adhere to the letter of the principle.

[11] James Rachels seems to urge that a restricted active euthanasia policy be instituted on the basis of, among other things, the favorable consequences. *See* his widely anthologized Active and passive euthanasia. *N. Engl. J. Med.* 292 (1975) 68–80. Compare Tom Beauchamp's response in 'A reply to Rachels on active and passive euthanasia, op. cit. Beauchamp, too, acknowledges some favorable consequences of instituting the policy, though he does not find them decisive.

[12] Cf. Tom Beauchamp, A reply to Rachels on active and passive euthanasia, op. cit. Beauchamp enumerates these and other consequences that he maintains are the reasonably expected results of instituting a policy permitting active euthanasia. The italicized phrase above is Beauchamp's; the emphasis is mine.

[13]The quotation marks are intended to convey the dubious assessment of the anticipated changes in attitudes. The assessment in the rule utilitarian argument is based on our current desires, attitudes, and dispositions, which are simply not relevant to a utilitarian assessment of future desires, attitudes, and dispositions.

Abstract

In the relevant circumstance of both relatively high burden or cost to others and low prospective gain for themselves, people have a prima facie *duty to die sooner rather than later. This duty is a personal moral duty, not a "societal" duty, and thus, it in no way implies that others may demand that it be exercised. It is literally a duty to let death come, not a more general duty to die that encompasses active measures to ensure death. A person can owe this duty not only to close family and loved ones when they are heavily burdened, but in the right circumstances also to more distant persons, even to "society"; the fundamental reason is fairness to others in the society or insurance pool, given one's own evaluative preferences about trade-offs involving expensive, low-chance-of-benefit care. This personal moral duty to die does not extend to incompetent patients, at least not in any literal sense. The duty to die also does not jeopardize the marital vow, "for better or worse, in sickness and in health"; the creative and mutual relationship envisioned in a marital promise does not bar people from ever coming to think that they may be obligated to accept death because of effects on a spouse. Finally, although this duty is rooted in sensitivity to considerations of quality of life, it need not threaten the "special" value of individual life and lifesaving.*

The Nature, Scope, and Implications of a Personal Moral Duty to Die

Paul T. Menzel

Introduction

Arguments for and against any claim that people can have a duty to die when they arrive at a certain appropriate combination of difficult circumstances need to pay careful attention to the nature and scope of the alleged duty. I will argue that in the relevant circumstance of both relatively high burden or cost to others and low prospective gain for themselves, people have a prima facie personal moral duty to die sooner rather than later. Unlike a recent defender of the duty to die, John Hardwig,[1] I defend a version of this duty that can be owed not only to close family and loved ones when they are heavily burdened, but in the right circumstances, it can also be owed to more distant persons, even to "society." In another respect, however, my interpretation of the duty to die is less extensive than Hardwig's: it does not extend to incompetent patients, at least not in any literal sense. Moreover, contrary to the views of Daniel Callahan, I will argue that the existence of this duty does not imply that others may demand that it be exercised, nor does it jeopardize the marital vow, "for better or worse, in sickness and in health." Finally, I argue that although this duty is rooted in sensitivity to considerations of quality of life, it need not threaten the special value of individual life and lifesaving.

Any alleged duty to die is undoubtedly going to be extremely controversial. Even if (by some miracle) my arguments are logically compelling, they will presumably be so within the limits of some set of other plausible claims and assumptions. They will therefore not appear compelling to those who do not share those assumptions. Since debate about whether people may at some point have a duty to die will certainly not remotely result in any societal consensus anytime soon, the most important thing to establish at this point in time is the legitimacy of the discussion itself. The case for such a duty to die is plausible and persuasive enough, and any dangers that it might create unjust pressures on the vulnerable are modest enough that we should fear neither its articulation in specific individual situations nor its open defense and discussion in the wider society.

In outlining the duty that I will defend, three things need to be emphasized initially.

1. It is only a *prima facie* duty. That is, even when its complex qualifying conditions are met and the duty is present, it can still be outweighed by countervailing moral forces that may prevail in particular circumstances.
2. It is only crudely stated as a "duty to die." More accurately, it is a "duty to let death come relatively cheaply." That is, it only comes into play in circumstances of considerable burden or cost to others, and since it can be performed by refusing to continue expensive life-extending measures, it need not require the use of active lethal agents.

3. It is a personal moral duty. As such, one cannot infer that it may be enforced by the action or pressure of others. Although the duty morally obligates those who have it, it does not by itself give others the right to enforce it or to blame people for their failure to perform it.

The first of these three primary characteristics can pass without further comment, but the second and third characteristics warrant further explanation.[2]

A Personal Moral Duty to Let Death Come

The duty that I would defend need only to be a duty to passive, not active, behavior. Literally, it is a duty to let death come, not more generally a duty to die. To be sure, a duty to die could be construed more broadly to encompass situations in which the only way to die and remove the burdens on others is by active lethal means. Although I myself do not find the alleged moral relevance of the line between active lethal means and refusal of life-sustaining measures sufficiently persuasive to prohibit all active means of dying, that is a separate issue, the argument over which need not enter the debate over a duty to die. If good moral reasons hold against using active lethal means, then those reasons can similarly set a limit to any duty to die.[3] Defending the duty to die need not push us into defending its more active form any more than permitting refusal of treatment requires us to permit active killing.[4]

The matter of personal moral duty warrants a much longer explanation. Moral claims range over a very wide spectrum of choice: from rules imposed on people without mediation by any choice or action of their own to entirely optional and supererogatory acts "beyond the call of duty." On the former end, in the context of health care, lie decisions to "ration" or deny services.

In those cases, a patient is not even asked, much less required, to do or decide anything—a service is simply denied. In such rationing contexts, we might plausibly say that people "ought to die," but that expression does not state an individual's obligation or duty. In the rationing context, after all, there is nothing for the person to do. Here "ought to die"—even "duty to die"—may be used, but they only convey a belief that the decision to ration was morally correct.

On the other end of the spectrum of choice lie actions "beyond the call of duty"—saintly and heroic deeds. Here individual choice is key; any attempt by others to pressure people to perform supererogatory deeds is out of place. To be sure, we might believe, especially in reference to our own behavior, that we "ought" to do what is the very best. In a societal context, however, we leave the performance of a truly supererogatory deed entirely up to the individual. Not only is one not required to do what one heroically ought to do, but, one should not suffer blame if one does not. That is simply the logic of supererogation.

In the middle of this spectrum fall three other kinds of moral claims: "societal" moral duties (including, in many cases, legal duties), "personal" moral duties, and cases in which we say something is "a good thing to do." Next to rationing, societal moral duties leave the least room for choice. They also leave the most room for social pressure. Demands may legitimately be made on people to perform these duties. Though "required" and "demanded," however, the actions they require still "run through" the persons holding the duty—they have to do something. If they do not live up to their duty, of course, they may reap the consequence of either legal punishment or social blame. In "personal" moral duties, on the other hand, people may strongly blame themselves if they do not perform their duty, but others may not properly demand that they carry out the obligation. Despite that, personal moral duties weigh much more heavily on people than prospective saintly or heroic deeds. When one is not saintly or heroic, one is not blamed, and when one does such a exemplary deed, one is singularly honored or praised. With personal moral duties, by contrast, one readily blames oneself for nonperformance, perhaps even very sternly (that is why we think of them as duties).

We may also discern an additional segment of the spectrum between personal moral duty and supererogation: "good things to do." They constitute less duty than personal moral duties, but are not so strictly optional as supererogation. Suppose, for example, that for several months, I forgo my frequent afternoon cappucino and with the savings buy a truly special gift for my spouse. I am hardly thereby a saint or a hero, but on the other hand, barring some particularly tough time recently in my spouse's life, I had no duty to do such a thing—not even a personal duty, much less a societal or legal one. Clearly, though, it was "a good thing to do."

The entire range from least to most choice of the individual then is comprised of:

1. Decisions imposed by others.
2. Societal moral obligations (including many legal obligations).
3. Personal moral duties.
4. Good things to do.
5. Saintly or heroic deeds.

Critics of a proposed duty to die sometimes fail to notice the distinctions among these different kinds of moral claims. In particular, they fail to notice the category of personal moral duty and its distinction from harder obligations. Daniel Callahan, for example, prognosticates that if I had a duty to die, others "would have the right to demand that I die."[5] They would indeed have such a right if the duty to die were a more societal or legal duty, but if the duty to die really is a personal moral duty, others will have no right to demand that I die. However, if there is a duty to die, is it a personal moral duty?

Take the comparison to the more "required" side—societal moral duty. Were the duty to die not a personal moral one, but a more societal duty, others would be justified in demanding performance. Substantive moral considerations, though, should precede and determine the category of duty with which we are dealing, not vice versa. Are others justified in demanding performance? If, all things considered, they are not, and if the moral claim is stronger than supererogation or "good deeds," then we may be looking precisely at the kind of moral claim that I have called "personal moral duty." To be sure, the substantive question of whether others are justified in demanding performance may be difficult to answer (and I will not pursue it further here, but only later in the conclusion). Nonetheless, an important point for our current discussion can still be made. If critics are right that others would not be justified in pressuring a person who allegedly has a duty to die by demanding performance, the duty to die does not wither away. It may still be alive, but only as a personal moral duty.

Take the alternative in the opposite direction on the spectrum of choice: is the alleged personal moral duty to die perhaps not a duty at all, but rather only a case of supererogation or "good deeds"? The duty to die, I would argue, obtains only where the burdens on others, financial or otherwise, are great and the benefits of further life (or the likelihood of life) are relatively small. Hardwig's paradigm case is one relevant type: an 87-year-old mother has a 50 percent chance of 6 more months of life if aggressive medical treatment is pursued, but if it is, and especially if it is successful, her 55-year-old caregiver daughter (the only remaining child) will lose virtually all her savings, her home, and likely her career. (I will describe another paradigm case in the next section.)

In cases like this, do we really believe that the appropriate moral claim is only that the mother would be doing something saintly or heroic if she refuses to continue aggressive treatment? When the stakes are modest for the person who has the opportunity to die while they are great for others, we should take the gloves off our moral language. Dying more cheaply and less expensively can be more than an admirable sacrifice; it can sometimes be morally required.[6] Just as we encourage moral vanity and distort moral reality if we speak of the person who only stops by the roadside or gives to Oxfam as a saint or hero, so we also corrupt ourselves if we think that refusing to create sizable burdens by chasing after a bit more life is heroism. A person may just be doing his or her duty.[7]

Can the Duty Be Owed to Society as Well as to Family?

John Hardwig limits the duty to die to the relatively close-knit contexts of family and loved ones. The burdens and costs that help generate the duty must fall on them, not on others more distant in society. Thus, the duty is owed only to them. His examples (one of which has been described above) as well as his arguments directly reflect this limited scope of his version of the duty to die. Within this scope, he can appeal poignantly to family relationships. There "the fact of interwoven lives debars us from making exclusively self-regarding decisions... The impact of my decisions upon my family and loved ones is the source of many of my strongest obligations and...the...likeliest basis of a duty to die." Hardwig dismisses, by contrast, the possibility of any duty to die based on burdens and costs to those more distant in society with only a few sentences. "'Society,' after all, is only very marginally affected by how I live, or by whether I live or die," and "I am not advocating a crass, quasi-economic conception of burdens and benefits, nor a shallow, hedonistic view of life."[8] For Hardwig, that concludes the matter.

Undoubtedly the case for a duty to die is made more easily—or is a stronger duty—in such close-knit contexts. Not only are the burdens much more highly concentrated on specific individuals, but most people believe that generally obligations are more numerous and stronger the closer the social circle. Partly this is owing to the very serious roles that parents accrue when they have (or decide to have) children, and partly it is owing to the sheer amount of interaction within most families.

Let me describe another case in a very different context, however.[9] Mr. Smith, 83 years old, develops acute pancreatitis. After several days, physicians estimate that with very aggressive treatment, he has a 75% chance of getting back to moderately good health, though long-run damage would undoubtedly limit the length of any ensuing recovery to under five years. Within 10 days, they predict, they will know whether things are turning around toward eventual recovery, or whether Mr. Smith's odds will have dropped to at best 1 in 10. Were the odds then to drop, he could be kept alive in the ICU for another 3 or 4 weeks at the cost of another $60,000–75,000. During virtually all of that time, he would be unconscious. Mr. Smith is aware of the outlines of this prognosis during his few days of competent consciousness in the early stage of the crisis. He consents clearly to the strategy of aggressive treatment for 10 days, but says nothing about what should be done thereafter if his odds of recovery sharply drop. Were they to drop, would his care to retain a 1-in-10 chance of meaningful recovery for at most another 5 years be worth the additional $60,000–75,000 expense (most of it insured)? Another 3 weeks of care in the ICU will not be easy on his family, but no careers, life savings, home ownerships, or family relationships would be threatened.

Does Mr. Smith have a duty to stipulate, now while he is still conscious and competent, that if the odds do turn down, he will be quickly allowed to die? Hardwig would say he has no such duty. No terribly poignant burdens parallel to the loss of a home or career are at stake for any family member. However, the man has lived a fulfilling 83 years, and the benefit to him of aggressive and expensive care after a downturn in the odds is relatively small. Translating his chances into a cost-per-statistical-benefit relationship, continued care in circumstances like his costs $120–150,000 per year of life saved. Were we generally to spend up to that level in the society, health care expenditures would constitute an even greater proportion of the nation's economic product than they do already. Moreover, Mr. Smith himself has contributed regularly to Habitat for Humanity, which builds houses for the needy for less than $75,000 apiece. In effect, then, the choice is whether he should consume the equivalent of a house before he very likely dies. I would argue that here it is nearly as plausible to argue that the benefits of continued aggressive treatment are too small in this case to compensate for the dispursed costs that others will have to pay as it is for Hardwig to argue in his paradigm case that the benefits of life-sustaining care for the mother are too small to compensate for the toll that care takes on the daughter.

The most powerful factor behind such an argument in the Smith case, however, is not a mere direct weighing of the benefits to the patient with the opportunity costs of the admittedly dispursed expenses to others. The most powerful factor focuses more on Mr. Smith himself: what would Mr. Smith himself have said were he to have been consulted about this $75,000 expenditure for a 0.1 chance of at most a 5-year recovery at the time he signed up to pay the insurance premiums?[10] In his other investments in safety and health care, he does not reflect nearly as high a value as $120,000 per statistical year of life saved. Therefore, we can quite comfortably predict that he would not have purchased an expansion of his insurance to cover such low-benefit/high-relative-expense hospital care. Is it not, then, inconsistent for him now to press ahead with such expensive care? Very likely Mr. Smith was not willing to pay premiums at a rate at which the insurer could afford to cover the expense of continuing his current care (and the care of others similarly situated). His duty to accept death when his odds of recovery sharply drop is therefore owed to his fellow insurance subscribers: it would not be fair to them for him now to demand provision of this expensive, low-chance-of-benefit care, for that would shift the extra costs to them without paying his proportionate share himself.

Such unfairness may typically justify not a personal moral duty, but a more imposed rationing of Mr. Smith's care. The larger society or insurance pool, however, may not yet have moved to this level of control over life-extending care. Perhaps that is even for good reason—our presumption of his unwillingness to pay insurance premiums at the rate necessary to support care at this cost–benefit level is fallible. Perhaps only he is in a position to know whether such a presumption is correct. In any case, if he would not have been willing to invest in his future care at the necessary rate of $120,000 per statistical year of life saved, it is unfair of him now to call on the insurance pool to cover aggressive care once the chances for his recovery plummet. Such unfairness grounds a stronger moral claim—personal moral duty—than merely the opportunity for supererogatory sacrifice.

Incompetent Patients

Cases like Mr. Smith's may often not lend themselves to the language of personal moral duty, or duty at all, because the patient becomes incompetent before reaching the opportunity to refuse treatment and die. Ironically, although Hardwig is opposed to extending the duty to die beyond the circle of family and close friends, he defends its extension to patients who have become mentally incompetent. He can "make no sense of the claim that someone has a duty to die if the person has never been able to understand moral obligation at all…, [but] those who were formerly capable of making moral decisions could have such a duty."[11]

This strains the logic of "duty." How can people have duties that they cannot decide to carry out? If they ought to die and that decision is justifiably imposed on them, that is imposed denial of care, not an implementation of their duty. Maybe others have a duty to impose that decision, but then what is left of the patient's duty to die? We do not have a good phrase to denote one person's duty to let another person die, but it is hardly perspicuous to speak of it as either person's "duty to die."

In any event, Hardwig's reason for extending the duty to incompetent patients is instructive. After briefly acknowledging the oddness of saying that an incompetent patient can have a duty to die, he notes that "very urgent practical matters turn" on the matter. "If a formerly competent person can no longer have a duty to die..., I believe that my obligation may be to die while I am still competent.... So I must die sooner than I otherwise would have to," but that is too great a sacrifice to allow a duty to die. Hardwig's solution is to say that the later, newly incompetent person can have a duty to die. Then, "if I could count on others to end my life after I become incompetent, I might be able to fulfill my responsibilities while also living out all my competent or semi-competent days."[12]

Indeed the last sentence is correct, but note: if I can count on others to end my life after I become incompetent, this would also block the duty to die from creating pressure on me to use active aid-in-dying before there is any life-sustaining care to be withdrawn or withheld. This moral situation is not correctly described as my duty to die. It would be a clearer use of language to say only that now incompetent persons can previously have had a duty to instruct others to let them die later, and others may then have a duty to let them die. However, they, as incompetent persons, do not now have a duty to die.

Marriage Vows

Daniel Callahan argues that at least within a family context, reliance on a benefit–burden calculus is tantamount to destroying—or at least jeopardizing—the marriage vow. In Callahan's view, Hardwig is "saying in effect: 'for better or worse, in sickness and in health—well, sort of, it all depends.'"[13] Is that what Hardwig and defenders of a duty to die that applies in family contexts are in effect saying? Whatever a duty to die does convey, does it represent a moral deterioration of marital and family vows?

What is the content of the promise "for better or worse, in sickness and in health"? Do people necessarily intend and understand it as absolutely barring consideration of burdens on the other that might obligate a very ill partner to accept death sooner rather than later? I doubt that the promise is that clear—in fact, it seems clear that it is not that clear. The vow charts a direction and certainly bars people from running from a marriage simply because their partner is sick and burdensome. Why, though, should we think that the promise itself is capable of addressing such weighty later issues as whether a short period of additional life for one partner can justify the sustained imposition of extremely destructive burdens on the other? The larger point of the vow is to put people on notice that simply following their self-interest is not acceptable; they will indeed be obligated to make sacrifices. Although those sacrifices will sometimes be very large, it is hardly clear on the face of the vow's language that they must be unlimited just because it is the partner's sickness and health that are involved. More importantly, the vows are explicitly taken mutually. Why would the creative and mutual relationship envisioned in a marital promise be seen as barring people from ever coming to think that they were obligated to accept death because of effects on a spouse? "In sickness and in health" hardly commits people to thinking that just any amount of life or health—or just any increase in the probability of retaining it—can outweigh other goods in their partners' lives.

To be compatible with a personal moral duty to die, therefore, this promise hardly needs to be reinterpreted as a promise to care for the other in sickness and in health "sort of, it all depends." A "sort of, it all depends" qualification of the vow sounds inappropriate, of course—colloquially, that is clear language for diminishment. But why should a vow that keeps open the largest questions about the most extreme potential circumstances be seen as diminished? "Sort of, it all depends" is not at all accurate language for the kind of potential context in which one partner might accrue a duty to die. Thus, it is not fair to accuse Hardwig or other advocates of a selective personal moral duty to die of advocating, in effect, the addition of such a phrase to the traditional vow.

Retaining Special Value for Life and Lifesaving

How destructive of a "special" value for life and lifesaving is a willingness to consider effects on quality of life? If one has a duty to die in situations where the remaining quality of life available is very low and others' quality of life is greatly affected by continuing to live, quality of life is influential on both sides of the judgment. If a patient's quality of life is very low, to the point where that person is even somewhat ambivalent about continuing to live, the very values of the patient in deeming his or her own life low in quality would seem to permit consideration of others' quality of life.

We might wonder whether the matter is perhaps different when we shift the patient's situation to one where the benefit of a patient's care appears to be small, not because of his or her diminished quality of life, but because of the low probability that his or her care will successfully extend life at all. This is part of Mr. Smith's scenario. When after 10 days of aggressive care his condition does not improve, the worry is not that continuing the care risks saving him for a severely compromised future condition. The relevant new fact is that although any continued aggressive care that did work to prolong his life 6 months, for example, would leave him in a state of substantial recovery, the odds that that care will extend his life remotely that long in any condition are slim. Can we still respect the value of life if we discount the benefits of care simply by the mathematical probability that they will come to pass?

One strategy of response for either of these situations is to trace the "special value" of life and lifesaving down to the hard choices that people are willing to make when presented with difficult trade-offs. Suppose that the people who are now patients are willing to trade some of their own time in a life of low quality for a sufficiently enhanced quality of life.[14] Suppose that they would also prefer an interpersonal trade-off in which a sufficiently large number of persons in a condition of low quality of life were cured before the very lives of a sufficiently smaller number were saved.[15] That is, they are rather robustly willing to sacrifice some life itself for enhanced quality of life. They prefer such a trade of sufficient proportions not only in an individualistic situation within their own lives, but also in a societal context in which trade-offs between people are at stake.[16]

Empirical research has so far discovered that people are distinctly cautious in trading life extension for some people off against curative improvements for others. The number of persons that people say must be cured by one program to be equivalent to another program that saves a certain number of different persons' lives is considerably greater than the necessary number that might be extrapolated from those same persons' more individualistic trade-offs within their own lives.[17] These preferences are indicative of what we might call the "special value" of life and lifesaving. They should lead us to employ very cautiously the argument that trade-offs between a patient's low benefit from continued life support and other persons' high costs and burdens actually represent the patient's own assent to a particular trade-off proportion. Such trade-offs may indeed represent such assent, but only in the proportions indicated by the more cautious interpersonal trade-off preferences that bridge between enhancing quality of life and extending life.

There is reason to think that similar, cautiously interpreted preference data can be used to ground an argument that when a person's probability of longer-term life extension drops sufficiently low, it can no longer justify imposing certain costs and burdens on others. If we discount the value of a chance of later life simply by its mathematical improbability, however, we may underestimate the value that people put on care that provides a low chance of success. Even reflective people tend to place a higher than proportionate value on treatments that harbor such chances.[18] We might call this the extra value that accrues to "maintenance of hope."[19]

A duty to die gains considerable justification from the very values expressed by the people who later accrue that duty in specific situations. That justification can apply both to cases where the benefit sacrificed by the person with the duty is time in a life of distinctly low quality and to cases where the benefit sacrificed is a relatively low chance of extending life at all. In articulating such justification, however, and in discerning how low the benefits sacrificed by patients and how high the burdens and costs relieved for others must be before the duty applies, one must be careful not to misinterpret and inflate people's willingness to trade off life for other values. Empirical evidence of a limited, but "special value" of life resides in two phenomena: people demand greater compensating gains before sacrificing life in interpersonal trade-off contexts, and they assign higher than mathematically proportionate ("maintenance of hope") value to treatments with low probability of success.

Conclusion

These and other subtleties about people's values may be one more reason for arguing that the duty to die ought to remain a personal moral duty within the discretion of the individual who has the duty, not a duty where others may demand or enforce performance. Such a personal moral duty, however, is hardly innocuous. Duties left socially and legally unenforced can still be influential without becoming insidiously coercive on the most vulnerable among us. The personal moral duty to die is an inevitable response to a situation in which death is no longer regarded as the greatest evil, but in which it is admittedly terribly difficult to determine when patient benefit is sufficiently low and costs and burdens to others sufficiently high. A culture can sensitize people to these moral possibilities if its members speak frankly of duties to die while leaving the task of actual discernment and performance of duty to those who will actually pay its price.

Notes and References

[1]Hardwig, John (1997) Is there a duty to die? *Hastings Center Report* 27, 2 (March–April): 34–42.

[2]These two attributes characterize the duty to die that I defend in pp. 190–203 in *Strong Medicine: The Ethical Rationing of Health Care*. New York: Oxford University Press, 1990. Part of the next section is a review of the discussion there.

[3]It is conceivable that some elements would be present in duty-to-die, but not right-to-die situations that would argue persuasively for extending the duty to die, but not the right to die to encompass active lethal means. Even were that the case, however, the duty to die would not rest or fall on how the active/passive debate turned out.

[4]Suppose, on the other hand, that permitting refusal of lifesaving treatment did logically push us relentlessly toward permitting active killing. Then, of course, the duty to let death come may indeed become a more active behavior encompassing duty to die. Precisely those who would object to such a duty, however, will deny the supposed premise. They, then, can hardly object that a duty to let death come ineluctably pushes us toward the broader duty to die.

[5]Callahan, Daniel (1997) Letter in response to Hardwig (note 1). *Hastings Center Report* 27, 6 (November–December): 4.

[6]Caution is clearly in order in claiming that this person has a duty to her daughter to die sooner rather than later. Neither Hardwig nor I claim this is more than a *prima facie* duty. If, for example, the mother had endured truly great sacrifice to prevent or significantly ameliorate some disaster to the daughter, or if certain genuine and voluntary promises had been made earlier in their lives, then perhaps the daughter's burdens do not finally create a duty to die.

[7]Most of the language in the last four sentences is taken directly from *Strong Medicine* op. cit., p. 195.

[8]Hardwig, op. cit., p. 36.

[9]The case is creatively constructed from a 1994 experience in my family.

[10]The moral force of such "presumed consent" is explored and defended at length in *Strong Medicine* op. cit., pp. 22–36.

[11]Hardwig op. cit., p. 39.

[12] Hardwig op. cit., p.39.

[13]Callahan op. cit.

[14]Health economists involved in discerning the relative value of different health states call this the "Time Trade-Off."

[15]Health economists call this the "Person Trade-Off" method of measuring the comparative value of different treatments.

[16]For the difference between the "individual health state utility" that is measured by methods, such as the Time Trade-Off, and the "societal value" of health improvements that is measured by the Person Trade-Off, *see* Nord, Erik (1995) The person trade-off approach to valuing health care programs. *Medical Decision Making* **15;** 201–208. On the different roles of these two kinds of value, *see* Nord, Erik, Pinto, Jose Luis, Richardson, Jeff, Menzel, Paul, and Ubel, Peter (1999) Incorporating concerns for fairness in numerical valuations of health programmes. *Health Economics* 8,1 (January), 25–39.

[17]One study reporting this effect is Nord, Erik, Richardson, Jeff, and Macarounas-Kirchman, K. (1993) Social evaluation of health care versus personal evaluation of health states: evidence on the validity of four health state scaling instruments using Norwegian and Australian surveys. *Int. J. Technol. Assess. Health Care* 9: 463–478. For normative analysis related to such empirical work, *see* Menzel, Paul, Gold, Marthe, Nord, Erik, Pinto Prades, Jose Luis, Richardson, Jeff, and Ubel, Peter (1999) Towards a broader view of values in cost-effectiveness analysis of health care. *Hastings Center Report* 29, 3 (May–June): section II-2.

[18]Ubel, Peter, and Loewenstein, George (1995) The efficacy and equity of retransplantation: an experimental survey of public attitudes. *Health Policy* 34: 145–151.

[19]The phenomenon receives this label in Menzel et al. op. cit.

Abstract

In this paper, I support Paul Menzel's argument that there is sometimes a moral duty to die in certain medical contexts. Then I argue that the moral duty to die might be extended to the contexts of criminal justice. In so doing, I juxtapose Immanuel Kant's claims that there is a duty not to commit suicide and that there is a duty for murderers to die. When the State is unable to impose capital punishment on murderers deserving of it, then such murderers are in no way relieved from their moral duty to die. They have a moral duty to die by way of suicide. Kant must then modify either his claim against suicide or his retributivist stance toward murderers. Finally, I analyze the moral duty to die, setting forth and defending the conditions under which, if satisfied, one has a moral duty to die.

Analyzing the Moral Duty to Die

J. Angelo Corlett

Introduction

In recent years, much has been made of the alleged moral right to die.[1] Whether it is derived from a cluster of fundamental moral interests/claims[2] to self-ownership, freedom, and/or equality,[3] a putative moral right to die has been the subject of many debates in philosophical circles. If there were such a right, then it would seem that it might in some circumstances imply a correlative moral duty[4] of others to refrain from interfering with the right holder's morally legitimate exercise or benefit of the moral right to die.

However, is there a moral *duty* to die?[5] If so, under what conditions might one have such a duty, and why? In this paper, I shall distinguish between different grounds for the putative moral duty to die. Then I shall analyze the conditions in which such a moral duty could exist. For my present purposes, "death" shall be taken to mean "the intentional death of a human being."[6]

The Moral Duty to Die Inexpensively in Medical Contexts

Paul Menzel has argued for a subtle (and beyond the purposes of this article, a more developed) version of the contractarian approach to the nature of the putative moral duty to die.[7] Given the realities of the "age of delayed degenerative diseases," and in light of the importance of considerations of the need to conserve valuable resources, we have a duty to die inexpensively.[8] This duty would be a special instance of the more general moral duty to die.[9] Menzel argues, "All of us would be better off in the long run if we would have agreed that sometimes even when there is still net value left in life, we should let people die."[10] He continues, "sacrificing the benefit of time in life might be attended by a compensating positive benefit of knowing that one is parting with a small share of life for the benefit of others."[11] Arguing that it is egoistic at times to cling to the preservation of life come what may, Menzel states:

> If preserving lives of declining quality in old age is much less a benefit to the aged patient than the resources saved can be for others, then it will be in the mutual self-interest of all to have a general practice of letting death come more efficiently.[12]

According to Menzel, then, it is egoistic for me to preserve my life when it is declining in quality if the cost of doing so would mean significant adverse effects to others. Indeed, I have a moral duty to die cheaply under such circumstances so that the economic and other resources saved by my cheap death could be used to enhance the quality of other lives. Moreover, Menzel avers, "dying more cheaply and less expensively is not just admirable sacrifice; sometimes it is morally required."[13]

However, precisely what is the moral duty to die cheaply on Menzel's account? Whatever else it is, it is grounded in moral agreement. Moreover, it is justified on the grounds that such a duty maximizes resources, presumably for society at large, including future generations.

It seems that Menzel's arguments, if plausible, support an imperfect moral duty to die cheaply, rather than a perfect one, for his analysis appears to allow some role for inclination to play in determining when one has such a duty. Furthermore, the right to die cheaply appears to be a positive moral duty.

If, as Menzel argues, there is sometimes a moral duty to die inexpensively, then it might be argued that such a duty accrues to a patient to the extent that medical costs are delimited to only what is morally justified, all things considered. In other words, there is a moral duty to die inexpensively to the extent that medical costs are made inexpensive relative to justifiable costs of, say, a particular procedure that is needed to sustain or end a patient's life. This claim is meant to follow from "'ought'[14] implies 'can.'"[15] There cannot be a moral duty to die inexpensively where medical costs to die are prohibitive. In a society where medical costs to die are affordable, then in certain circumstances, a moral agent has a perfect moral duty to die, but where such a condition does not prevail, the moral duty to die, if it accrues at all, cannot be a strict one. For there can only be a moral duty to die cheaply where dying cheaply is possible, all things considered.

Therefore, if there is sometimes a moral duty for persons to die inexpensively, then it seems that there is a moral duty of society to ensure that medical costs are as inexpensive as good quality of health care (for either sustaining or ending life) will permit. Here we ought not assume that the more expensive the medical treatment, the higher its quality, or that the less expensive it is, the lower its quality. Moreover, if medical costs are extravagant, then both living and dying in medical contexts are overly costly, and this would appear to make impossible (for practical purposes) the imperfect moral duty to die cheaply. Thus, Menzel's arguments in favor of the moral duty to sometimes die inexpensively in medical contexts implies a moral duty to make medical costs of dying affordable. As Margaret P. Battin argues, "a redistributive policy cannot be just without adequate guarantees that resources will in fact be redistributed as required."[16]

There are various ways in which the costs of health care for dying might be made more affordable. One way to reduce such costs is to reduce the costs of medical technology and research. Another is to reduce the costs of medical/hospital administration. Yet another is to reduce the costs of physicians' services. I shall focus on the latter two prospects.

Menzel argues that:

> At a time when talk of cost containment is common, little serious and comprehensive literature has been produced on the normative *justification* of physicians' incomes, as opposed to the mere description of what they are or the explanation of what causes them to be that way...
>
> ... If life, lifesaving, and health have a finite price, doctors' earnings are less likely to be worth our money. *Who gets paid what* is just as crucial an allocation question as how many of our resources are used on health care, and for whom. Asking the most difficult and basic questions about the cost of health care will inevitably lead us to a discussion of physicians' incomes.[17]

What Menzel argues concerning physicians' salaries[18] might also hold for medical/hospital administrators and their respective incomes, for based on a plausible principle of justice as distributive equality,[19] physicians' incomes are not justified to be as high as they are, generally, whether or not they are based in the free market. Thus, there is a kind of logical coherence between Menzel's major contributions to medical ethics, and if his arguments are plausible, then the moral duty to die sometimes cheaply grounds the moral duty to make medical costs of living and dying affordable. This in turn implies that physicians' and medical/hospital administrators' incomes be set objectively at only that mark that is morally justified, all things considered.

Now it might be argued that there is no moral duty to reduce medical costs, contrary to Menzel, and that lacking such a duty, there is no moral duty to die inexpensively. Medical providers, physicians, and so forth, deserve, the argument goes, whatever they can get, come what may. After all, the argument continues, if " 'ought' implies 'can'" is plausible, then there is no moral duty to die cheaply.

However, what grounds this *laissez-faire* argument in medical contexts? If the argument is not to fall prey to a moral arbitrariness objection, it must endorse, it would appear, a more general *laissez-faire* view of costs of living and dying in a society, costs that take us beyond medical contexts. However, this more general point might well be refuted by a plausible version of social contract theory, whether Thomas Hobbes'[20] or another.[21] Thus, the *laissez-faire* argument for unrestricted health care costs seems problematic or is at least in need of adequate rational support.

There appear, then, to be good reasons to infer that medical costs require delimitation and in multifarious ways. These reasons include the problematic nature of a *laissez-faire* system of politics and economics (especially concerning its implication for medical costs), not to mention the plausibility of there being a moral duty to die affordably in certain contexts. This makes room for Menzel's arguments for reduced physicians' salaries and for my extension of his point to medical/hospital administrators' incomes as well. Given that "'ought' implies 'can,'" there is good reason to think that there is a moral duty sometimes to die inexpensively, as Menzel argues. Also, it is precisely this sort of argument that seems to serve as a foundation for our attempting to discern death's meaning in even the most vexing circumstances. Perhaps an element of death's meaning would be that when I choose to die inexpensively, I would not sacrifice the futures of my loved ones or of others who might benefit from my dying expensively. Altruistic citizenship is more meaningful than egoistic self-fulfillment, even if self-interest might at times override one's duty to die inexpensively for the sake of others.

The Moral Duty to Die in Criminal Justice Contexts

The putative moral duty to die has been discussed philosophically and primarily in terms of specified medical contexts, but the moral duty to die also accrues, if it accrues at all, in some contexts of criminal justice. Therefore, for one to have a moral duty to die implies, among other things, that one ought to act or refrain from acting in a particular way, under certain circumstances, such that the end result of one's act or omission is that one is dead, whether by one's own "hand" or by another's. If there were such a duty, then it would imply that society has a *prima facie* moral obligation to refrain from interfering with the agent's exercise of his or her duty to die. Thus, if a criminal has a duty to die, for reasons of, say, social utility or desert, then society and the legal system have a *prima facie* moral obligation to not interfere with the agent's duty to die.

The moral duty to die might accrue in contexts of retributive justice (on a Kantian account of retributive punishment), where I commit an act so untoward and deserving of my death that I am, morally speaking, required to die. This would imply that no amount of forgiveness or mercy[22] would suffice to neutralize my duty to die. In fact, it might be argued that the State has under such conditions a duty not to interfere with my duty to die. If I have a duty to die based on my criminal "mind"/conduct, which makes me deserving of death, this assumes that I satisfy strongly the conditions of criminal and moral liability. Given these conditions, the State has a duty to punish me by death. This duty is based on my breaking rules in a manner that makes me morally deserving of death such that my death is not a moral prerogative,[23] but a *requirement.*[24]

Furthermore, contrary to Kant's absolute prohibition against suicide (i.e., his claim that one has a perfect moral duty to oneself to not commit suicide), it might be argued that there are instances in which, on Kantian retributivist grounds, one has a moral duty to commit suicide. Such a case might be borrowed from Kant's *The Metaphysical Elements of Justice, 333,* where he argues that a society seeking to disband itself must first put to death all of those rightly convicted of murder.[25]

However, suppose a murderer knows he or she has murdered and has strongly satisfied, while murdering, the conditions of criminal liability. Clearly, that person deserves to die (on Kantian grounds) and knows it. But what if the State cannot put him or her to death? Perhaps the State is experiencing political revolution or secession, making its ability to administer retributive justice practically impossible for an indeterminate amount of time. Does not the murderer have, on Kantian grounds, a perfect moral duty of virtue to commit suicide under these circumstances? Or alternatively, what if the State decides itself not to put to death this murderer? Could it not, for all Kantian justice says (short of pardoning him or her), sentence him or her to suicidal capital punishment? Moreover, does not the murderer under such circumstances have a perfect moral duty of virtue to commit suicide? After all, Kantian retributive justice requires that all murderers receive a death sentence (recall Kant's own words regarding murderers: "If, however, he has committed a murder, he *must* die"[26]) unless there are significant mitigating circumstances that would dictate otherwise. Also, what grounds such proportional punishment is the notion of moral desert. Therefore, such cases, the murderer deserves death, and if the State cannot or will not administer capital punishment on a murderer deserving of it, then this surely would not, for all Kant says, relieve the murderer himself or herself of the perfect moral duty of retributive justice to commit suicide in order to effect on himself or herself the "appropriate" or proportional punishment. Although this point is not logically entailed by Kant's view of punishment, it is consistent with his position on punishment to say that murderers deserving of death have a perfect moral duty to die, whether by the hand of the State, or by some other means, including suicide.

Assumed here are the Socratic claims that:

1. (Some forms of) injustice and wrongdoing (such as murder) are (at least among) the greatest of evils;
2. Punishment rids the wrongdoer of the evils of her wrongdoings; and especially
3. Wrongdoers ought to seek their own (appropriate or proportional) punishment for their wrongdoings "to prevent the distemper of evil from becoming ingrained and producing a festering and incurable ulcer in his soul."[27]

If this reasoning holds true, then Kant's claim that we have a perfect moral duty of virtue to not commit suicide is questionable.

Therefore, there might be a moral duty to die based on a rule the breaking of which means the rule-breaker deserves to die, and there are a number of other possible grounds for the moral duty to die, including utilitarian, contractarian, and other such grounds.[28] To be sure, the duty to die might be grounded in a hybrid of such views.

Analyzing the Moral Duty to Die

What are the conditions in which I have a moral duty to die? It might be argued that I have no moral duty to die unless I, being a moral agent, have a capacity to know, intend, and act freely, all of which are necessary for my having a moral duty to die. I shall refer to this as the "Agency Condition." Although some might think that the Agency Condition is unnecessary in that it is precisely a human who lacks such agency capacities who has a moral duty to die, it seems to be true of moral duties in general that the Agency Condition is required for duty-bearers. It is, moreover, the Agency Condition that makes it difficult for moral duties to accrue to nonhuman animals. It would seem, then, that this capacity is a necessary condition of my having a duty to die, but what else is necessary, and what is sufficient for my possessing such a moral duty?

It might be argued that if " 'ought' implies 'can,'" I must be able to die at a given time and in a given place if I indeed have a moral duty to die at that time and in that place. I shall refer to this as the "Ability Condition." Thus, I have no moral duty to die, and I am unable to die. If I am, at a particular time and place, unable to effect my own death, directly or indirectly, or if there is no other means for me to die, at that time and in that place, then I have no duty to die then and there. Another or I must be able to somehow cause my death, either directly or indirectly, for me to have a moral duty to die. However, this condition, if satisfied, is not sufficient for me to have a moral duty to die. Simply because I am able to die, inexpensively or not, this in itself does not morally require me to die.

Additionally, I have a moral duty to die, expensively or not, by some means if either I deserve to die because, say, I have caused sufficient unwarranted harm to others to justify my death, or because I, in remaining alive via medical assistance, would cause sufficient unwarranted harm to others such as others' pain or impoverishment. I shall call this the "Moral Liability Condition." However, this is only a necessary condition for my having a moral duty to die, given that " 'ought' implies 'can.'" There might well be circumstances in which my death cannot, practically speaking, be intentionally caused even though I deserve to die.

Collectively, the Agency Condition, the Ability Condition, and the Moral Liability Condition amount to a cluster of criteria that when satisfied, indicate a moral duty to die. I proffer them as jointly sufficient conditions of the moral duty to die.[29]

An Objection and Replies

It might be argued that the above analysis of the moral duty to die ignores a crucial factor about morality and humans. The fact is, the objection goes, humans have intrinsic worth, and anything resembling a putative moral duty to die is always trumped by a fundamental and absolute right to life shared by every moral agent. The very notion, then, of a moral duty to die is nonsense, whether in medical or nonmedical contexts. I shall refer to this as the "Intrinsic Moral Worth Objection."

In reply to the Intrinsic Moral Worth Objection, it might be argued that it not only begs the moral question concerning whether or not there can ever be a moral duty to die, but it also fails to see that the human right to life is not absolute, for there are various circumstances in which we would have no right to life, such as if we are the on the morally "wrong side" of a war, or as the Moral Liability Condition has it, if we ought to die based on either utilitarian or retributivist reasons. Thus, the Intrinsic Moral Worth Objection seems to be problematic, for it appears that its reason for the absolute moral right to life of each moral agent is neutralized by considerations in favor of the moral duty to die.

Furthermore, the Intrinsic Moral Worth Objection to the above analysis of the moral duty to die appears to be contingent for its overall plausibility on there being a plausible notion of intrinsic moral value. After all, it is humans who are said to have intrinsic moral worth, value in and of themselves, according to the objection. However intuitively appealing the concept of intrinsic moral value is, there are some important questions raised concerning its plausibility.[30] Thus, there is reason to doubt the potency of the Intrinsic Moral Worth Objection as a defeater of the above analysis of the moral duty to die.

Conclusion

Having set forth a basic notion of a moral duty, I articulated the nature of a moral duty to die according to deontological, intuitionist, utilitarian, and social contract standpoints. Then I argued for a particular analysis of the moral duty to die, defending it against an important objection. Perhaps this analysis will better enable lawmakers to devise legal rules that will enjoy the support of objectively true moral principles that will in turn better enable us to understand under what conditions each of us ought to choose death over life as a matter of moral character and integrity.

Acknowledgments

I am grateful to Robert Almeder, Peter Atterton, Steve Barbone, Robert Francescotti, James Humber, Paul Menzel, and Mark Wheeler for their helpful comments on an earlier draft of this paper.

Notes and References

[1]For some recent research on this topic *see* Margaret P. Battin (1994) *The Least Worst Death.* Oxford University Press, New York; *Ethical Issues in Suicide.* Prentice-Hall, Englewood Cliffs, 1995; John Keown, ed. (1997) *Euthanasia Examined.* Cambridge University Press, Cambridge.

[2]For philosophical analyses of the various views of rights, *see* Ronald Dworkin (1977) *Taking Rights Seriously.* Harvard University Press, Cambridge; Joel Feinberg (1980) *Rights, Justice, and the Bounds of Liberty.* Princeton University Press, Princeton; *Freedom and Fulfillment.* Princeton University Press, Princeton, 1992, Chapters 8–10; Wesley Hohfeld (1919) *Fundamental Legal Conceptions.* Yale University Press, New Haven; Loren Lomasky (1987) *Persons, Rights, and the Moral Community.* Oxford University Press, Oxford; Robert Nozick (1974) *Anarchy, State, and Utopia.* Basic Books, New York; Joseph Raz (1986) *The Morality of Freedom.* Clarendon Press, Oxford; L. W. Sumner (1987) *The Moral Foundation of Rights.* Oxford University Press, Oxford; Judith Jarvis Thomson (1990) *The Realm of Rights.* Harvard

University Press, Cambridge; Jeremy Waldron (1993) *Liberal Rights.* Cambridge University Press, Cambridge; Carl Wellman (1985) *A Theory of Rights.* Rowman & Littlefield,Totowa; *Real Rights.* Oxford University Press, New York, 1995; *The Proliferation of Rights.* Westview Press, Boulder, 1999.

[3]For philosophical analyses and discussions of these concepts, *see* G. A. Cohen (1995) *Self-Ownership, Freedom and Equality.* Cambridge University Press, Cambridge. For a critical discussion of Cohen's analyses of these and related concepts, *see J. Ethics* 2 (1998): 1–98.

[4]For a discussion of the correlation between rights and duties, *see* David Lyons (1970) The correlativity of rights and duties. *Nous* 4: 45–55; On a generous interpretation, a version of the "weak correlativity thesis" about rights and duties is found in F. H. Bradley (1951) *Ethical Studies.* Bobbs-Merrill, Indianapolis; p. 144. Furthermore, not all duties are correlated with rights, as Joel Feinberg notes (Duties, rights, and claims, in *Rights, Justice, and the Bounds of Liberty*, p. 139):

> ...duties of indebtedness, commitment, reparation, need-fulfillment, and reciprocation are necessarily correlated with other people's *in personam* rights. Duties of respect and community membership are necessarily correlated with other people's *in rem* rights, negative in the case of duties of respect, positive in the case of duties of community membership. Finally, duties of status, duties of obedience, and duties of compelling appropriateness are not necessarily correlated with other people's rights.

Feinberg's denial of a strict correlation between moral and legal duties and rights is also found in Joel Feinberg, Human duties and animal rights in *Rights, Justice, and the Bounds of Liberty,* p. 186.

[5]Throughout this chapter, I do not distinguish between duties and obligations, though some rather subtle distinctions are made between them in Richard B. Brandt (1964) The concepts of obligation and duty. *Mind* 73: 374–393; E. J. Lemmon (1962) Moral dilemmas. *Philos. Rev.* 71: 139–158.

[6]For discussions of the concept of death, *see* Fred Feldman (1992) *Confrontations with the Reaper.* Oxford University Press, Oxford; Some puzzles about the evil of death. *Philos. Rev.* C (1991): 205–272; Thomas Nagel (1979) *Mortal Questions.* Cambridge University Press, Cambridge, Chapter 1; Ingmar Persson (1995) What is mysterious about death? *South. J. Philosophy,* XXXIII: 499–508; Palle Yourgrau (1987) The dead. *J. Philosophy,* LXXXIV: 84–101.

[7]Paul Menzel (1990) *Strong Medicine.* Oxford University Press, Oxford, Chapter 11.

[8]Menzel, *Strong Medicine,* Chapter 11.

[9]Menzel articulates and defends a conception of the duty to die inexpensively in medical contexts. Such a duty differs, however, from the duty to die generally in at least the following ways. First, Menzel argues for a personal moral duty to die cheaply, one that, unlike the more general moral duty to die, is not exercised at all by enforcement activities of others (such as when the State punishes criminals). The person, moreover, who has a moral duty to die cheaply desires still to live, whereas those who have a more general moral duty to die need not desire to live.

[10]Menzel, *Strong Medicine*, 193.

[11]Menzel, *Strong Medicine,* 194.

[12]Menzel, *Strong Medicine*, 194.

[13]Menzel, *Strong Medicine,* 195.

[14]By "ought" I mean a word expressing moral obligation, just one possible meaning of the term, according to Gilbert Harman (1977) *The Nature of Morality.* Oxford University Press, Oxford, pp. 84–87.

[15]James W. Corman, Keith Lehrer, and George S. Pappas (1992) *Philosophical Problems and Arguments,* 4th ed. Hackett, Indianapolis, p. 304. For more on "ought" implies "can," *see* Toni Vogel Carey (1985) What conflict of duty is not. *Pacific Philos. Q.* 66: 204–215; Charles F. Keilkopf (1967) "Ought" does not imply "can". *Theoria*; Walter Sinnott-Armstrong (1989) "Ought" conversationally implies "can". *Philos. R.* XCIII: 249–61.

[16]Margaret P. Battin (1987) Age rationing and the just distribution of health care: is there a duty to die? *Ethics* 97: 340.

[17]Paul Menzel (1983) *Medical Costs, Moral Choices.* Yale University Press, New Haven, p. 214.

[18]Menzel, *Medical Costs, Moral Choices,* pp. 224–226. Menzel is careful to focus his discussion on a specific type of physician, namely certain specialists whom he believes are overpaid for a variety of reasons. However, I believe that Menzel's arguments concerning the unjustifiability of certain physicians' salaries are sound and apply to physicians more generally. The same arguments, I believe, apply in general to medical/hospital administrators.

[19]Perhaps consistent with the egalitarianism articulated and defended in G. A. Cohen (1991) Incentives, inequality, and community. *The Tanner Lectures on Human Values*. Stanford University, 21, May 23; The Pareto argument for inequality. *Soc. Philosophy Policy,* 12 (1995): 160–185.

[20]Gregory Kavka (1986) *Hobbesian Moral and Political Theory.* Princeton University Press, Princeton.

[21]David Gauthier (1986) *Morals by Agreement.* Oxford University Press, Oxford: *Moral Dealing.* Cornell University Press, Ithaca, 1990; John Rawls (1971) *A Theory of Justice*. Harvard University Press, Cambridge; *Political Liberalism.* Columbia University Press, New York, 1993. For commentaries on Rawls' version of contractarianism, *see* J. Angelo Corlett, ed. (1991) *Equality and Liberty: Analyzing Rawls and Nozick.* Macmillan and St. Martin's, London and New York.

[22]Jeffrie G. Murphy and Jean Hampton (1988) *Forgiveness and Mercy.* Cambridge University Press, Cambridge.

[23]For a discussion of agent-centered moral prerogatives and agent-centered moral requirements, *see* Samuel Scheffler (1982) *The Rejection of Consequentialism.* Oxford University Press, Oxford; Samuel Scheffler, ed. (1988) *Consequentialism and Its Critics.* Oxford University Press, Oxford.

[24]This view seems to be consistent with the analysis of punishment found in Immanuel Kant (1965) *The Metaphysical Elements of Justice,* John Ladd, trans. MacMillan, New York, p.102. For discussions of Kant's view of retributive punishment, *see* Jeffrie G. Murphy (1987) Does Kant have a theory of punishment? *Columbia Law Rev*. 87; J. Angelo Corlett (1993) Foundations of a Kantian theory of punishment. *South. J. Philosophy*, XXXI: 263–284.

[25]Immanuel Kant, *The Metaphysical Elements of Justice*, p. 102.

[26]Kant, *The Metaphysical Elements of Justice*, p. 333.

[27]Plato, *Gorgias,* 479d–480a.

[28]I discuss such positions in "Is there a moral duty to die?" forthcoming.

[29]My knowing that I have a moral duty to die (what I shall refer to as the "Epistemic Condition") is neither a necessary nor a sufficient condition of my having such a duty. It is not necessary because I can indeed possess the duty, but not know it, and it is not sufficient because my knowing that I have such a duty will not ground it, if I, say, cannot die by other than "natural causes" at a particular time and place.

[30]*See* Shelley Kagan (1998) Rethinking intrinsic value. *J. Ethics* 2: 277–297; Thomas Hurka (1998) Two kinds of organic unity. *J. Ethics* 2: 299–320; Noah Lemos (1998) Organic unities. *J. Ethics* 2: 321–337; Fred Feldman (1998) Hyperventilating about intrinsic value. *J. Ethics* 2:, 339–354; Robert Audi (1998) The axiology of moral experience. *J. Ethics* 2: 355–375.

Abstract

John Hardwig created a major stir in the bioethics and health care ethics communities when he published his provocative article "Is There a Duty to Die?" In response to this article, Rosemarie Tong argues that individuals have no duty—either to family members, other loved ones, or society—to die.

First, Tong summarizes Hardwig's arguments on behalf of a duty to die, particularly his view of this duty as a moral obligation or responsibility as opposed to a legal duty. Second, she outlines five points in the case against a duty to die:

1. *It seems inappropriate to hinge the duty to die on whether or not a person is fortunate enough to have a family, loved ones, or another type of web of human relationships.*
2. *Given our present society, in which health care expenses and managed care regulations influence the giving of less medical treatment instead of the giving of all possible treatments, considerations of a duty to die are inappropriate until standards are set for managed care treatment, for example.*
3. *A duty to die will not be perceived equally by the sexes and therefore may be discriminatory toward women who may feel pressured to accept a duty to die, because men are typically socialized to think of their individual rights and women are typically socialized to think of their communal responsibilities.*
4. *Instead of a "duty" to die, the concept might be better understood as the "option" to die, a choice that people are free to make.*

5. *Instead of using the language of duty and obligation, the concept should be described in the language of caring and choice.*

Finally, Tong concludes that in our present individualistic and rights-oriented society, it is not safe to posit a duty to die. Until all members of society are considered on equal terms, the duty to die will not be imposed on citizens equally.

Duty to Die

Rosemarie Tong

John Hardwig created a major stir in the bioethics and health care ethics communities when he published his provocative article "Is There a Duty to Die?"[1] For persons more accustomed to asking questions about a purported right to die,[2] Hardwig's question had an immediately destabilizing effect. Why, they wondered, is there a need to propose a duty to die when so many family members and health care providers routinely fail to follow the instructions in patients' advance directives—instructions that specify wishes not to be aggressively treated or kept "alive" under certain circumstances?[3] Are there, they further wondered, a host of patients somewhere so desperate to cling to life that someone like Hardwig has to remind them of their duty to die—that enough is, after all, enough?

Like defenders of the right to die, I too was puzzled by the urgent tone of Hardwig's article, but for somewhat different reasons. Why, I reflected, do bioethicists insist on emphasizing individuals' rights and duties with respect to living and dying? Are life and death really matters of entitlement and obligation, or are they realities whose moral meaning transcends the limited and legalistic language of rights and duties? Leaving it for others to decide whether we have a right to die, what I intend to argue is that we have no duty to die. More specifically, I hope to persuade readers that deciding to die from a motive of duty or obligation undermines precisely the kind of human connections Hardwig wishes to strengthen. In order to make my case, however, I will first need to summarize Hardwig's arguments on behalf of a duty to die, for his arguments are, to my way of thinking at least, a particularly strong defense of this purported obligation.[4]

The Case for a Duty to Die

In presenting his case on behalf of a duty to die, Hardwig stresses that this duty is not a legal duty, but some sort of moral "obligation" or "responsibility."[5] Citing the example of Captain Oates, a member of Admiral Scott's expedition to the South Pole, who sacrificed his life so that his fellow explorers could survive, Hardwig notes that most people believe that there is some sort of duty to die, but that it is "uncommon."[6] He then observes that this uncommon duty might soon become a common one. As he sees it, our technological ability to prolong life permits too many of us to live too long in a debilitated state, draining those we claim to love of their vital energies and material resources. Therefore, in the very near future, we might be compelled to acknowledge that "[t]here can be a duty to die before one's illnesses would cause death, even if treated only with palliative measures. In fact, there may be a fairly common responsibility to end one's life in the absence of any terminal illness at all. Finally, there can be a duty to die even when one would prefer to live."[7]

Clearly, Hardwig takes the duty to die seriously. He implies that a relatively happy 55-year-old man who is beginning the descent into the depths of Alzheimer's, Huntington's, or Parkinson's might very well have a duty to phone Dr. Kevorkian with a request for some immediate "aid-in-dying."[8] Such a man, suggests Hardwig, not only might, but will have such a duty if his continuing existence threatens to harm other individuals seriously, particularly those individuals to whom he is most closely related. Hardwig argues that bioethicists typically reject a duty to die because their thought processes are held captive to what Hardwig terms "the individualistic fantasy,"[9] according to which our lives are separate and unconnected. From my point of view, Hardwig rightly concludes that such a self cannot exist for "we are not a race of hermits."[10] At all times, but especially when we are most vulnerable in body, mind, and/or spirit, we are dependent on others, most usually our family members, and, it is the fact of our dependency on others that is probably the source of many of our strongest obligations to others.

Seeking support for his claim that among our strongest obligations to others is a duty to die for our family members and loved ones' sake, Hardwig stresses that most people report that they do not want to burden their intimates, and that, in particular, they do not want to drain them of their physical, emotional, and financial resources. That most people express such feelings does not surprise Hardwig, for as he sees it "those of us with families and loved ones always have a duty not to make selfish or self-centered decisions about our lives."[11] In words that would make utilitarians beam, Hardwig announces that individuals blessed with families and loved ones should always "choose in light of what is best for all considered."[12] Therefore, if my intimates view caring for my disintegrating self as a tremendous benefit rather than an enormous burden to them, then "I have no duty to die based on burdens to them,"[13] but if they view me as depriving them of a normal life-style, for example, then I probably do owe them a duty to die. To be sure, concedes Hardwig, families and loved ones owe me a certain level of caregiving, but I must always remember that the cross I bear is fundamentally mine—it is meant for me—and that in determining whether I owe it to others to die, my interests and preferences are no more important than theirs.

Convinced that the only real argument against a purported duty to die is the negative effects one's dying might conceivably have on others, Hardwig readily dismisses three other arguments against this duty. Observing that we live in a fundamentally secular society, he immediately rejects the argument that it is God's prerogative to determine the moment of an individual's death. Next, Hardwig discounts the view that in deciding to die to ease others' lives, an individual "disrespects" his or her own life, implying that it is not worth as much as other persons' lives. On the contrary, insists Hardwig, an individual disrespects himself or herself by cowardly clinging to life instead of courageously offering it to those whom he or she claims to love the most in this world. Finally, Hardwig counters the point that it is cruel to burden already vulnerable, dying people with a duty to die with the counterpoint that it is equally cruel to burden those who care for the dying with a duty to minister to them no matter what. After all, it is the dying person's caregivers and not the dying person who will have to live with the long-term consequences of the dying person's decision to keep clinging to life: the depleted finances, the physical and psychological scars, and so on.[14]

Repeatedly, Hardwig stresses that death is not the worst event that can befall a human being, a point that Socrates made centuries ago when he chose to drink poison hemlock rather than to stop teaching Athenian youth to live an "examined" life—i.e., a meaningful life truly worth living.[15] There is, says Hardwig, something noble about the duty to die, a duty that he then qualifies as neither universally applicable nor unilaterally determined by one's self. Whether one has such a duty will, he says, depend on a host of very particular and contextual circumstances, but especially to the degree one's continuing existence is experienced as burdensome by one's caregivers. Relatedly, one's ability to perform this duty will depend on whether or not one is competent, for incompetent persons cannot, for example, depend on others to execute the terms of their advance care directives to do their duty for them.[16] Thus, if Hardwig is right, only an individual who is:

1. Competent;
2. Enmeshed in a network of caregivers who find his or her continuing existence exceedingly burdensome; and
3. A member of a society whose publicly funded, long-term care programs are so stingy that families and friends are literally forced to exhaust themselves caring for their loved ones,

will have a golden opportunity to die a meaningful, self-sacrificial death. Other individuals, not so "blessed," will have to find another way to exhibit their virtue.

The Case Against a Duty To Die

As someone who, like Hardwig, eschews the "individualistic fantasy," why is it that I regard the above state of affairs as morally odd—indeed, morally perverse? Reflecting on my intuitive rejection of a duty to die, there seem to be several reasons accounting for its intensity. In order of increasing seriousness, they are as follows. First, it seems inappropriate to hinge the duty to die on whether one is fortunate enough to be enmeshed in a thick web of meaningful human relationships. Second, announcing a duty to die in a society whose paradigm end-of-life case study is no longer the Karen Quinlan case (a "right to die" case), but the Helga Wanglie case (a "right to live" case) is worrisome. Third, because of the way men and women are typically socialized in our culture, with men learning to think in terms of their individual rights as separate selves and women learning to think in terms of their communal responsibilities as connected selves, women are more likely than men to respond positively to the duty to die. Fourth, what Hardwig calls a "duty to die" might be better described and understood as what philosopher J. O. Urmson calls a "work of supererogation."[17] Fifth, and finally, if Hardwig espouses the views he says he has about the self–other relationship, he should probably abandon the language of duty and obligation altogether, substituting for it the language of caring and choice.

Enmeshment in a Web of Human Relationships as the Context for the Duty to Die's Emergence

In her book entitled *Women and Evil,*[18] Nel Noddings discusses the topic of euthanasia. Although her focus is primarily on the dying person's pain and suffering, and his or her "right" to call an end to it, Noddings views the euthanasia decision as a collaborative decision the dying person should make together with his or her family members, friends, and even health care providers. Specifically, she writes that "when someone is suffering and wishes to die, all affected parties should take part in the decision. A support group consisting of physicians, nurses, family, patient, and perhaps a trained advocate of sorts could talk with each other about what is best for the patient and for everyone else who is suffering."[19] Noddings implies that if, for example, a patient's family members and friends would find his or her decision to die too distressing, then that patient may be under some sort of obligation to go on living, but if a patient's support group turns out to be less than supportive—if, for example, "unloving relatives... suggest strongly to an elderly person that he or she has a duty to die and be done with it"[20]—then that expression of cruelty may be "sufficient reason"[21] for a patient to choose death. Why go on living, if no one loves me, implies Noddings.

Although Noddings and Hardwig both want the decision to die to be a collective decision, and although there is growing evidence that typically people (particularly women) wish to involve their family members and loved ones in their end-of-life decisions,[22] Noddings, unlike Hardwig, confronts the cruelty factor head-on. In the face of some caregivers' cruelty, the question then becomes: Do care-receivers really owe a duty to die to caregivers who weight their own interests far more heavily than the interests of those for whom they supposedly care? Also, more specifically, do dying or nearly dying parents owe their very lives to extremely ungrateful adult children, or are there some decisions and actions they owe no one, not even an extremely grateful adult child? I do not think my older son owes it to me not to marry a woman whom I and everyone else in the family regards as a "troublemaker" bound to wreck her havoc on us willy-nilly, nor do I think my younger son owes it to me to make wise decisions about his life-style, health care, and future career simply because he knows if he gets in a "tight spot" I will bail him out if it is in any way possible for me to do so. Why, then, should I think that my 85-year-old mother-in-law, who is exhausting our resources, owes it to me and my husband to die sooner rather than later? Is it not wrong for us, in the name of human closeness, to expect our loved ones to serve the collective, familial good when it comes to deeply personal decisions such as choice of mate, life-style, health care, career, or time of death? Also, if it is not wrong to make such demands on our loved ones, then it seems better—at least in certain circumstances—to be without loved ones and therefore free to decide one's fundamental fate for one's self. In sum, my choice seems to be between leading a quality family life, which generates duties like the duty to die, or leading a lonely life, but remaining free of such heavy duties.

From Karen Quinlan to Helga Wanglie: Shifting Society's Focus from the Right to Die to the Duty to Die

The paramount medical ethics case of the 1970s was the Karen Quinlan case, which resulted in the articulation of patients' right to refuse treatment, including life-sustaining treatment.[23] Not Quinlan's health care providers, but Quinlan's family wished to remove the respirator that assisted her ability to breathe. Her family and friends repeatedly made the point that Karen would not want to live the "life" of a person in a persistent vegetative state, and that she would prefer death to an unconscious life that permitted her to experience neither knowledge nor love nor activity. When the United States Supreme Court finally ruled in the Quinlans' favor, it was to serve what they identified as Karen's "liberty interest" to be free of unwanted medical support.[24] Commenting on the Court's decision, one reporter observed: "The framers of the Constitution, who prohibited the Government from depriving a person of 'life' or 'liberty' without due process of law, had little reason to envision a day when the very act of sustaining a life might itself be a deprivation of liberty."[25]

The Quinlan's court victory ushered in a dramatic patients' rights movement. Gradually, citizens began to proclaim not only their right to refuse treatment, but also their right to die.[26] Furthermore, although the decision in the subsequent Nancy Cruzan case—which made it clear that incompetent patients with adequate documentation had the right to refuse all medical treatment, including artificial food and hydration— did not articulate a clear right to die, it certainly contributed to an increasingly favorable view on active, voluntary euthanasia and physician-assisted suicide, as well as on passive, voluntary euthanasia.[27]

Significantly, both the Quinlan and Cruzan decisions, in 1976 and 1990, respectively, were made in the era of fee-for-service medicine, which had among its faults a tendency to overtreat some patients with medically unnecessary and sometimes very costly surgeries and drugs.[28] Moreover, and particularly at the time of the Quinlan decision, American medicine was proud that it did not have to ration health care—that it gave the "best" health care to all patients, no matter what. Thus, if dying or nearly dying patients had disagreements with their health care providers it was with respect to refusing their bounty—i.e., saying "no" to surgeries and drugs they did not want, even if such refusals meant their death.

By the early and mid-1990s, however, patients were less inclined to fear overtreatment than undertreatment. Talk of medically unnecessary or futile medical care was everywhere, as the health care community debated whether the term "futility" should be interpreted primarily from a quantitative perspective (as inefficacious therapy with only a small percentage change of success) or instead primarily from a qualitative perspective (as therapy that permits the patient to survive, but "requires the patient's entire preoccupation with intensive medical treatment, to the extent that s/he cannot achieve any other life goods"[29]). Additionally, emphases on the high costs of health care, costs that society supposedly could no longer bear, were prevalent.[30,31] Finally, analyses of the emerging managed care industry led to the speculation that health care providers were being economically rewarded for limiting medical care, including medically necessary, nonfutile medical care.[32] Into this cauldron was dropped the case of Helga Wanglie.

Mrs. Wanglie was an 86-year-old Minnesota woman who fell and broke her hip in December 1989. Successfully treated at one hospital, she suffered respiratory failure as she was being transferred to another hospital in January. At that time, she was still cognizant of her family. When moved again, however, in May, Mrs. Wanglie had a heart attack and suffered brain damage. From that point on, she was unconscious. Moved back to the second hospital, she remained dependent on a respirator and tube feedings. Mrs. Wanglie was eventually diagnosed as a persistent vegetative state (PVS) patient, a medical judgment that led her physicians to conclude that further aggressive treatment would not benefit her. Neither the respirator nor her tube feedings would restore her normal bodily functions or enable her to experience the simple fact that she was alive. The Wanglie family objected on the grounds that their loved one's religious and moral convictions were such that she would benefit from further aggressive treatments. Life, they asserted, is intrinsically valuable and worth living under any and all circumstances, including the circumstances of persisting in a vegetative state. Faced with this degree of family opposition, the hospital's medical staff tried to get the district court to appoint an independent conservator to determine whether further aggressive treatment of Mrs. Wanglie was medically appropriate. The court decided, however, that an independent conservator was not needed, since no one could better identify and articulate Mrs. Wanglie's needs and interests than her husband of 53 years. Mrs. Wanglie died a few days after this decision, her ventilator and tubes notwithstanding.

Notably, Mrs. Wanglie's physicians maintained throughout Mrs. Wanglie's decline that further aggressive treatment was not in her best interests and that moreover, they were not serving any of medicine's traditional ends and aims[33] by keeping her alive. They never implied that Mrs. Wanglie had a duty to die, just that it was best for her to die. Critics speculated, however, that had Mrs. Wanglie not had a generous private health care insurance policy, which fully covered all of her considerable expenses, the duty to die might have emerged—in her case, a duty owed not to her family and loved ones, but to society in general.[34] For this reason, Susan Wolf among other bioethicists has argued that:

> To institute physician-assisted suicide and euthanasia at this point in this country—in which many millions are denied the resources to cope with serious illness, in which pain relief and palliative care are by all accounts woefully mishandled, and in which we have a long way to go to make proclaimed rights to refuse life-sustaining treatment and to use advance directives working realities in clinical settings—seems, at the very least, to be premature. Were we actually to fix those other problems, we have no idea what demand would remain for these more drastic practices and in what categories of patients. We know, for example, that the remaining category is likely to include very few, if any, patients in pain, once inappropriate fears of addiction, reluctance to sedate to unconsciousness, and confusion over the principle of double effect are overcome.[35]

Wolf, let it be known, was expressing her concerns about what she regards as a worrisome right to die movement—concerns that would, no doubt, increase were Hardwig's and others' arguments to inspire a duty to die movement.

Male/Female Differences with Respect to Rights and Duties

Another matter that Hardwig does not attend to sufficiently in his defense of a duty to die is the way in which a person's gender might affect his or her adherence to a duty to die. The differences between males and females in determining rights and duties has been explored for many ethical issues. For example, in her path-breaking book *In a Different Voice,* Carol Gilligan offers an account of women's moral development that challenges Lawrence Kohlberg's account of what he describes as human beings' moral development.[36] According to Kohlberg, moral development is a six-stage, progressive process. Children's moral development, he says, begins with Stage One, "the punishment and obedience orientation"; this is a time when to avoid the "stick" of punishment and/or to receive the "carrot" of reward, children do as they are told. Stage Two is "the instrumental relativist orientation." Based on a limited principle of reciprocity—"You scratch my back and I'll scratch yours"—children in this stage do what satisfies their own needs and occasionally the needs of others. Stage Three is "the interpersonal concordance or 'good boy-nice girl' orientation," when immature adolescents conform to prevailing mores simply because they seek the approval of other people. Stage Four is "the law and order orientation," when mature adolescents do their duty in order to be recognized as honorable as well as pleasant people. Stage Five is "the social contract legalistic orientation," when adults adopt an essentially utilitarian perspective according to which they are permitted considerable liberty provided that they refrain from harming other people. Finally, Stage Six is "the universal ethical principle orientation," when some, though by no means all, adults adopt an essentially Kantian perspective universal enough to serve as a critique of any conventional morality, including that of their own society. These individuals are no longer governed by self-interest, the opinions of others, or the rule of law, but rather, are motivated by self-legislated and self-imposed universal principles, such as those of justice, reciprocity, and respect for the dignity of human beings as intrinsically valuable persons.[37]

Puzzled by the fact that according to Kohlberg and his scale, women rarely climbed past the good boy-nice girl stage, whereas men routinely ascended at least to the social contract stage, Gilligan hypothesized that the deficiency lay not in women, but in Kohlberg's scale. She alleged his scale provided a way to measure not human moral development as he claimed, but only male moral development; she further explained that if women were measured on a scale sensitive to women's style of moral reasoning, women would prove just as morally developed as men.[38] With this hypothesis in mind, Gilligan conducted a study of 29 women who were in the process of deciding whether it was right or wrong for them to have an abortion. The more she spoke with these women, after as well as during their decision-making process, the more Gilligan realized that as a group, these women referred little to their rights and much to their relationships. In addition, she noted that in the process of developing (or failing to develop) as moral decision makers, these women typically moved in and out of three understandings of the self–other relationship, the third of which struck Gilligan as most morally developed:

1. An overemphasis on self.
2. An overemphasis on the other.
3. A proper emphasis on self in relation to the other.[39]

On the basis of this and several other empirical studies of women's moral reasoning patterns, Gilligan concluded that for a variety of cultural reasons, women in our society typically utilize an ethics of care, which stresses relationships and responsibilities, whereas men typically employ an ethics of justice, which stresses rules and rights.[40]

Even though Gilligan has conceded to her critics that not all women are care-focused and not all men are justice-focused, and that women and men in other societies exhibit styles of moral reasoning different from the ones typical in our society,[41] she has not entirely forsaken her gender-based claims over the years. For example, in one of her recent studies involving 80 educationally privileged North American adults and adolescents, Gilligan notes that although two-thirds of her men and women subjects raised considerations of both justice and care, they all tended to focus on one or another of these moral values, in particular. Whereas just as many women preferred justice over care as care over justice, only one man expressed a preference for the language of care over the language of justice. For this reason, Gilligan concludes that even if justice has largely lost its connection to the "masculine" (i.e., to men), care still retains its strong connection to the "feminine" (i.e., to women).[42] In other words, even though women are increasingly comfortable speaking the "hard" language of rights, particularly in the public sphere, men are not increasingly comfortable speaking the "soft" language of relationships there. Care continues to be a value associated with women rather than men, a "feminine" rather than a "masculine" virtue.[43]

Assuming that Gilligan's analysis is with merit, and that society as a whole as well as women in particular view women as more caring, giving, nurturing, and self-sacrificing than men, it is reasonable to hypothesize that the duty to die might rapidly be "feminized." Women might perceive themselves as imposing unreasonable burdens on others simply because they typically live longer than men, suffer from more chronic conditions than men,[44] and require more long-term care than men. Currently, women constitute 60% of the recipients of informal care[45] and an even higher percentage of formal care (three-fourths of nursing home residents aged 65 or older are female, and at ages 85 and over, one in four women—compared to one in seven men—resides in a nursing home).[46] Additionally, women's caregivers—themselves predominantly women (mothers, wives, daughters, and sisters in the private realm, and female health care professionals and female health care aides in the public domain)—might expect the female recipients of their care to do the "womanly" thing, i.e., the "dying-so-that-others-might-live-happily-ever-after" thing.

Although there is little research on the role patients' gender plays in the attitudes female health care practitioners have toward their long-term care and/or dying patients, there is considerable evidence that many female familial caregivers vacillate between what Gilligan identified as Level One and Level Two moral reasoning, before they settle into Level Two moral reasoning. Initially, women who provide in-home care for elders may, as Barbara Logue writes, "experience a strong sense of burden, feelings of resentment, inadequacy, guilt, and fears about their own present and future well-being."[47] Required to give up their time, energy, leisure activities, meaningful jobs, and ordinary routines, many a caregiver will focus on herself, bemoaning her lot as an "unfair" one. However, the more such a woman develops Level Two patterns of moral reasoning, the more horrified by her lack of love and devotion she may become, a feeling that may cause her to atone for her "sins" by putting yet more of herself into her caregiving tasks. When such a woman herself becomes needful of health care, memories of her "own negative or ambivalent feelings, role conflict, or even a sense of altruism"[48] will probably increase her desire not to be a burden to others. Wanting to think of herself as a good woman—that is, an undemanding woman who makes life as easy as possible for others—she will feel called to exercise the right or duty to die sooner than a man who has not been socialized to think that his right to life is any less important than someone else's right to life.

Understandings of Duty

Years ago I read and was most impressed by J. O. Urmson's essay, "Saints and Heroes."[49] Urmson points out that one common way to institutionalize moral concepts is to model all moral principles and virtues on those familiar legal and institutional rules that aim to control conduct by dividing all actions into three categories: the required, the permitted, and the forbidden. Required actions are duties, and these are the only actions that have moral merit; permitted actions are merely "all right"; and forbidden actions are wrong. Urmson elaborates that this widely accepted threefold classification does not capture certain moral distinctions we routinely make. In particular, there is no room in it for heroic and saintly deeds, defined as "those actions which may present themselves as duties to those who perform them, but are in fact far in excess of anything that any moral rules could plausibly require."[50]

According to Joel Feinberg, the historical interest of Urmson's essay consists of the fact that "he has mustered hard-headed, secular, 'utilitarian' reasons in support of a doctrine which has long had a place in Roman Catholic teachings and which was quite explicitly rejected by [the Protestant reformer Martin] Luther."[51] The Roman Catholic view is that there is a distinction between God's Old Testament commandments and Christ's New Testament counsels. Whereas the former must be obeyed under penalty of sin, the latter are optional advisements for those who wish to lead not simply morally adequate, but morally perfect lives. Interestingly, and importantly for the arguments being developed here, Feinberg notes that the medieval Catholic theologian Aquinas discussed Christ's counsels of perfection under three main headings: poverty, chastity, and obedience. Since Aquinas regarded the counsels of perfection as moral "add-ons," he believed that acting in accordance with them was not necessary for salvation.[52] The Protestant reformer Luther regarded Aquinas' view of the counsels as fundamentally flawed, however. There is not, he insisted, one morality for people with a low moral quotient and another morality for those with a high moral quotient. Everyone has the same moral quotient and is called to the same level of moral development. Whatever other talents God may have unevenly distributed, moral talent is not among them.[53]

Clearly, Hardwig falls into Luther's camp in his discussion of a duty to die, whereas I fall into that of Aquinas. I tend to think that like physical brawn, intellectual acumen, and emotional sensitivity, moral fiber admits of degrees—call them the "abnormal," "normal," and "optimal," if you wish. Thus, like Urmson, I believe that a two-tier system of morality makes a great deal of moral sense, with morally average persons required to be less virtuous, for example, than morally gifted individuals. We do not all have to be saints like Mother Teresa to pass through the gates of heaven, but we all do need to at least do no harm. Thus, I believe that the duty to die is less of a commandment than a counsel of perfection—a call that some people (the saintly and heroic types) believe is morally required of them in particular — and not a bona fide commandment or true duty, which is incumbent upon all individuals in general.

Having expressed my preference for Urmson's categorization of moral actions into the forbidden, permitted, required, and supererogatory, let me stress that I believe that an even better approach to the topic of duty is one that avoids the legalistic language of duties and rights altogether. In her book *Lesbian Ethics*, Sarah Lucia Hoagland challenges the concept of duty. She claims that "a focus on obligation and duty is part of the whole dominant-subordinate value of the rule of the fathers, which would have us substitute control for integrity."[54] Hoagland stresses that the ethics of duty has been typically viewed as superior to the ethics of care. "The idea," she says, "is that if we act from duty, we will do the 'right' thing even when we are not so inclined: for example, we will visit our friend on a cold winter day when we'd rather stay home in bed."[55] However, insists Hoagland, if it is really our friend in the hospital, we should not need duty to motivate us to visit her. We will visit our friend unless there is good reason for us not to, i.e., a reason that goes beyond our dislike of getting out on a cold winter day.

Further reflecting on appeals to duty, Hoagland notes that according to the Kantian tradition, the idea is not only that if we act from duty, we will always do the right thing, but also that if we are dutiful, we will have the right reason to do the right thing. Thus, if I sacrifice my life for my family, I must do so because I understand it is my duty to do so and not simply because of my strong feelings of love for them—feelings that Kantians view as ephemeral.

Now, it strikes me that unlike Kant, Hardwig wishes to give feelings an honored place in his moral system. He repeatedly emphasizes how much he loves his family and how much this love motivates his understanding of the duty to die. Thus, when Hardwig uses the language of duty, he uses it not because he believes (in accordance with Kant) that reason should reign supreme in our moral life, but because he wishes to address considerations about duty that Hoagland perceives as arising out of a utilitarian rather than a Kantian moral tradition. As Hoagland sees it, utilitarians appeal to duty in order to rank the importance of individuals' competing interests. She comments: "My friend has a need—being visited at the hospital—which competes significantly with my desire to stay home in my warm bed. Thus, I need a sense of obligation to sort it out, to help me be impartial in assessing needs, to show me that in this case I ought to put aside my own needs and interests in order to accommodate hers."[56] Similarly, Hardwig implies that his desire to go on living may come into conflict with his family's desire not to devote a large measure of their physical, emotional, and financial resources to his care. Should this turn out to be the case, Hardwig is confident that his sense of duty will help him decide whether his interest or that of his family is of greater moral significance.

Interestingly, and once again very importantly, Hoagland rejects utilitarian as well as Kantian understandings of duty. She maintains that both of these interpretations of obligation come from a failed moral tradition whose tenets have led Westerners down false moral paths for centuries. Hoagland articulates these destructive views as follows: "(1) that interacting is essentially a matter of sorting out competing claims, (2) that choice, therefore, is a matter of sacrifice and compromise rather than focus and creation, and (3) that caring is a matter of happenstance and at the very least needs the guidance of authority (reason) and duty to be worthwhile."[57]

In explaining the differences between caring and duty, and ultimately emphasizing that caring is more important than duty, Hoagland analyzes duty:

> The idea, from anglo-european tradition, is that acting from duty means we have a desire to do the morally right thing *because* it is the morally right thing, not just because we happen to care. As a result of the anglo-european split between reasoning and emotions, caring is viewed as a matter of mere

> happenstance. On this theory, when we care, we are simply reacting to whatever is around us. We have made no judgments in caring. Thus, we need a sense of duty to help us to do the "right" thing... Acting from duty and obligation relegates caring to the backgrounds of our interactions, or even renders it irrelevant.[58]

Refusing to split reason from the emotions, Hoagland views caring not as an ephemeral inclination, but as a deliberate choice without which deep connection between humans is impossible. She notes that the idea of duty cannot inspire us to love each other, to delight in each other, but only to avoid stepping on each others' toes. In contrast, the idea of caring can inspire us to become one—to step into each other's shoes and feel comfortable there.

Given the availability and power of the language of human caring, of vulnerable people joining together, which for me represents an improvement over the language of saints and heroes as well as the language of duty for duty's sake—two rather "male" languages in the final analysis—I continue to be puzzled by Hardwig's predilection for the language of duty. If I choose to die to ease your burdens, it will be not because I owe this action to you, but because I want to give it to you. Moreover, if I make this choice, I will make it not because society has shaped me into a stereotypic mom—a woman who gives to others until she has no more to give—but because I want to do something meaningful with my life.

To be sure, at this point in time, there are dangers with using the language of care, most of which have to do with the ways in which that language has been and still is associated with women. According to philosopher Bill Puka, for example, care can be interpreted in two ways: (1) in Gilligan's way, "as a general orientation toward moral problems (interpersonal problems) and a track of moral development,"[59] or (2) in his way, "as a sexist service orientation, prominent in the patriarchal socialization, social conventions, and roles of many cultures."[60]

If Puka's interpretation of care is correct, we must liberate care from its oppressive contexts, so that women and men (indeed, anyone who is less powerful than someone else) can care freely, without fearing that their caring for others will cause them not to care adequately for themselves.[61] Before we start dying for others, then, it would be a good thing to make certain, as Sheila Mullett writes, that we are caring nondistortively.[62]

According to Mullett, nondistortive care has the following characteristics:

1. Fulfills the one caring;
2. Calls on the unique and particular individuality of the one caring;
3. Is not produced by a person in a role because of gender, with one gender engaging in nurturing behavior and the other engaging in instrumental behavior;
4. Is reciprocated with caring, and not merely with the satisfaction of seeing the ones cared for flourishing and pursuing other projects;
5. Takes place within the framework of consciousness-raising practice and conversation.[63]

If Mullett's analysis of nondistortive care is on target, then it is best that we think long and hard before we start exhorting each other to die to make each other's lives somehow better. In the kind of society in which we live, a society that is neither particularly just nor truly caring, the duty to die will be distributed inequitably and sometimes with a measure of cruelty.

Although someone like Hardwig may, as he implies, be fortunate enough to be a member of a family in which one can care freely, sad to say his experience is not typical. Before we start talking about the duty to die, we need to be sure ours is a just society in which, for example, health care benefits and services are distributed equitably and, also importantly, in which the task of care giving does not fall on the shoulders of women far more than it does on men. To speak about a duty to die in, for example, a Confucian culture in which people are constituted and related through a set of reciprocal duties makes eminent sense. However, to speak about such a duty in our individualistic, rights-oriented society strikes me as perilous. Although, I agree with Hardwig that we would probably be a better society if we possessed a delicately calibrated set of reciprocal duties, the fact of the matter is that this kind of moral grid is absent in our society. Therefore, my duty to die might be interpreted by some as your right to see me dead. Thus, I think that, for now, we best focus on developing the kind of society in which it is safe to posit a duty to die. Until and unless it is fair to require the duty to die from all members of our society, I prefer to view dying as a choice, a gift I may give to others not so much because I am a saint or a hero, but simply because I love some people so much and they me that my dying is the choice we each wish to make.

Notes and References

[1]John Hardwig (1997) Is there a duty to die? *Hastings Center Report* 27, no. 2: 34–42.

[2]Margaret Pabst Battin (1995) *Ethical Issues in Suicide.* Prentice Hall, Englewood Cliffs, NJ, pp.180–197.

[3]Gregory E. Pence, *Classic Cases in Medical Ethics: Accounts of Cases That Have Shaped Medical Ethics, with Philosophical, Legal, and Historical Backgrounds*, 2nd ed. McGraw-Hill, New York, p. 31.

[4]Hardwig, p. 34.

[5]Ibid.

[6]Ibid.

[7]Ibid., p. 35.

[8]Ibid., p. 39,40.

[9]Ibid.

[10]Ibid.

[11]Ibid., p. 36.

[12]Ibid.

[13]Ibid.

[14]Ibid., p. 37.

[15]Kenneth L. Vaux (1992) *Death Ethics: Religious and Cultural Values in Prolonging and Ending Life.* Trinity Press International, Philadelphia, pp. 5, 168.

[16]Hardwig, Is there a duty to die? p. 39.

[17]J. O. Urmson (1970) Saints and heroes, in *Moral Concepts*, Joel Feinberg, ed. Oxford University Press, London, pp. 60–73.

[18]Nel Noddings (1989) *Women and Evil.* University of California Press, Berkeley.

[19]Ibid., p. 132.

[20]Ibid., p. 134.

[21]Ibid.

[22]Elizabeth Morrow (1987) Attitudes of women from vulnerable populations toward physician-assisted death: a qualitative approach. *J. Clin. Ethics* 8, no. 3: 287.

[23]Pence, *Classic Cases in Medical Ethics: Accounts of Cases That Have Shaped Medical Ethics, with Philosophical, Legal, and Historical Backgrounds,* 2nd ed., pp. 8–17.

[24]For the text of this decision, *see In the Matter of Karen Quinlan,* vol. II. University Publications of America, Arlington, VA, 1976.

[25]Linda Greenhouse (1990) Right to reject life. *The New York Times* June 27, A1.

[26]Derek Humphrey and Ann Wickett (1988), *The Right to Die: Understanding Euthanasia.* Harper and Row, New York.

[27]Dan W. Brock (1996) Borderline cases of morally justified taking of life in medicine, in *Intending Death: The Ethics of Assisted Suicide and Euthanasia.* Prentice Hall, Upper Saddle River, NJ, pp. 131–149.

[28]E. Haavi Morreim (1991) *Balancing Act: The New Medical Ethics of Medicine's New Economics* Klower Academic Publishers, Dordrecht, pp. 9–22.

[29]L. Schneiderman, et al. (1990) Medical futility: its meaning and ethical implications. *Ann. Int. Med.* 112: 949–954.

[30]Frank Marsh and Mark Yarborough (1990) *Medicine and Money: A Study of the Role of Beneficence in Health Care Cost Containment.* Greenwood, New York.

[31]John La Puma (1998) *Managed Care Ethics.* Hatherleigh Press, New York, pp. 31–42.

[32]Rosemarie Tong (1995) Towards a just, courageous, and honest resolution of the futility debate. *J. Med. Philosophy* 20: 169,170.

[33]Stephen H. Miles (1992) Medical futility. *Law, Med. Health Care* 20: 310.

[34]George D. Lundberg. American health care system management objectives. *J. Am. Med. Assoc.* 269: 2554,2555.

[35]Susan M. Wolf (1996) Gender, feminism, and death: physician-assisted suicide and euthanasia in *Feminism & Bioethics: Beyond Reproduction,* Susan M. Wolf, ed. Oxford University Press, New York, pp. 305,306.

[36]Carol Gilligan (1982) *In a Different Voice.* Harvard University Press, Cambridge.

[37]Lawrence Kohlberg (1971) From is to ought: how to commit the naturalistic fallacy and get away with it in the study of moral development in *Cognitive Development and Epistemology* T. Mischel, ed. Academic Press, New York, pp. 164, 165.

[38]Gilligan, p. 173.

[39]Ibid.

[40]Ibid., p. 151–174.

[41]Carol Gilligan (1987) *Moral orientation and moral development* in Eva Feder Kittay and Diana T. Meyers, eds. *Women and Moral Theory,* Rowman and Littlefield, Totowa, NJ, p. 25.

[42]Ibid.

[43]Rosemarie Tong (1993) *Feminine and Feminist Ethics.* Wadsworth Press, Belmont, CA, pp. 1–12.

[44]Barbara J. Logue (1991) Taking charge: death control as an emergent women's issue. *Women & Health,* 17, 4:103.

[45]Ibid.

[46]Ibid., p. 107.

[47]Ibid., pp. 104,105.

[48]Ibid., p. 105.

[49]J. O. Urmson, "Saints and Heroes," pp. 60–73.

[50]Joel Feinberg (1970) Introduction, in *Moral Concepts,* Joel Feinberg, ed. Oxford University Press, London, pp. 9,10.

[51]Ibid., p. 10.

[52]Ibid.

[53]Ibid.

[54]Sarah Lucia Hoagland (1988) *Lesbian Ethics: Toward New Value.* Institute of Lesbian Studies, Palo Alto, CA, p. 273.

[55]Ibid.

[56]Ibid., p. 274.

[57]Ibid.

[58]Ibid., pp. 277,278.

[59]Bill Puka (1990) The liberation of caring: a different voice for Gilligan's *In a Different Voice*. *Hypatia* 5, no. 1: p. 59.

[60]Ibid., p. 60.

[61]Ibid.

[62]Sheila Mullett (1989) Shifting perspectives: a new Approach to ethics, in *Feminist Perspectives,* Lorraine Code, Sheila Mullett, and Christine Overall, eds. University of Toronto Press, Toronto, p. 119.

[63]Ibid., pp. 119,120.

Abstract

John Hardwig has argued for the claim that in a variety of circumstances, a patient has a duty to end his or her life. His argument depends on the (largely correct) contention that putting off death will result in significant economic, physical, and emotional turmoil for the patient's family. Hardwig's argument, it seems, invokes the principle that a duty to a is incurred if refraining from a imposes a serious burden on others, particularly one's family. I contend, in opposition to Hardwig, that obligations or duties do not arise in this way. No one would seriously suggest, for instance, that I am required to die even if a family member stands to lose his or her life's savings, because in part I persist in living rather than succumb to death. After all, a family member has something to say about how his or her money is to be spent and, in general, he or she is under no requirement to squander his or her savings on a doomed relative.

How Could There Be a Duty to Die?

David Drebushenko

Professor John Hardwig writes movingly, but not altogether convincingly, of a patient's duty to die. His principal argument for claiming that patients have a duty to die is based on the fact that one ought not to subject one's family to financial, physical, and emotional ruin. In many cases, Hardwig contends, the probability of extending life is often nearly negligible and the considerable costs to a person's family do not justify abdicating a duty to die. Moreover, not only is the probability that life can be extended comparatively low, but neither is life significantly enhanced; this besets both the patient, who in the interest of putting off the inevitable is willing to endure even more misery, and family members, who must provide continuing emotional, physical, and financial support.

I have so far understated the claim(s) Hardwig is advancing. He writes:

> I certainly believe that there is a duty to refuse life prolonging treatment. But a duty to die can go well beyond that. There can be a duty to die before one's illnesses would cause death, even if treated with palliative measures. In fact, there may be a fairly common responsibility to end one's life *in the absence of any terminal illness at all.* Finally, there can

> be a duty to die when one would prefer to live. Granted, many of the conditions that can generate a duty to die also seriously undermine the quality of life. Some prefer not to live under such conditions. But even those who want to live can face the duty to die. These will clearly be the most controversial and troubling cases; I will, accordingly, focus my reflections on them.[1]

This passage is worthy of a closer look, for it contains a number of assertions that may disturb some and, at the very least, surprise and possibly shock others. Hardwig is claiming, among other things, that:

1. In certain instances, there can be a duty to refuse life-sustaining treatment.
2. Under some conditions, one can be required to die prior to having been overtaken by illness, and perhaps most striking of all;
3. On occasion, there is a duty to die even though no threat to life is posed by the presence of a terminal illness.

It would seem that on his view the "troubling cases" involve a duty to die when there exists a preference for continued life, but this is, seemingly, an odd position; I should think that a preference for an outcome that is at odds with what one is required by duty to do is common. Such a feature is what makes doing one's duty difficult.

In the opening pages of his paper, Hardwig pauses to ask why contemporary bioethics has not considered the question of whether there exists a duty to die. His answer is that the field is influenced by what he terms, "the individualistic fantasy." The fantasy amounts to the (mistaken) notion that we lead insular and unconnected lives, apart from others, including ostensible friends, acquaintances, and even, to some degree, our families. According to Hardwig, this fantasy prevents us from seriously entertaining the notion that we may have a duty to die, and, since this is the case, we are unaware that we ourselves may be under an obligation to end our lives when the time comes. Although I agree with Hardwig about the content of the fantasy as well as its prevalence in our society (in fact, the fantasy is distressingly common in this culture; individual rights though highly touted are, in certain respects, subject to abuse), I do not agree with him about the explanation. Hardwig seems to have overlooked what is, arguably, an equally plausible alternative explanation: contemporary bioethicists, moral theoreticians, and well-informed lay persons have not seriously considered the proposition that there is a duty to die because there is no such thing, or so I will urge.

Hardwig goes on to make the point that in spite of the prevalence of the individualistic fantasy, and quite in spite of the way the fantasy is reinforced by a complicated mix consisting, in my view, of jingoism, errant libertarianism, and just plain bad political theory, our lives are interwoven in countless ways. Our actions do touch others, though we may not always be aware of that fact even when our actions have a significant impact on the welfare and well-being of those closest to us. The persons most likely to be deeply and directly affected by our decisions and choices are often immediate family members. What makes this relationship especially complicated is the fact that we are connected in so many ways to these individuals. One way in which we are very intimately tied to such persons is through emotional bonding that has occurred over a span of time lasting for many years and, in some cases, stretching over generations.

It is here, with the connection to one's family, that Hardwig begins to make his case for a duty to die. He writes:

> But the fact of deeply interwoven lives debars us from making exclusively self regarding decisions, as the decisions of one member of a family may dramatically affect the lives of all the rest. The impact of my decisions upon my family and loved ones is the source of many of my strongest obligations and also the most plausible and likeliest basis of a duty to die. "Society," after all, is only very marginally affected by how I live, or by whether I live or die.[2]

Hardwig is rightly concerned with the prospect that at some point, one may run the risk of becoming a burden to one's family should one opt to pursue life rather than comply with what, in his view, is a duty. There are many ways in which, assuming a choice to continue life, one can tax one's family. As Hardwig notes, precisely because of our emotional involvement with these individuals, heavy tolls can be exacted from family members when we attempt to put off death. The possibility for emotional suffering cannot be overstated. Our anxiety over our failing health, not to mention the raw fear that arises from facing death, can be quite debilitating to those who have to witness such upheaval, but fear and anxiety, though felt chiefly by the patient, are not the only the emotions to be concerned about. Our relationships with members of our family can be damaged by other kinds of emotional difficulties. Persons who once loved and adored us can, with enough stress, come to regard us with feelings that run in the opposite direction. In addition, there is very often a concomitant feeling of remorse, guilt, or even depression that a family member may experience because he or she has fallen out of love and into something altogether unsavory. Quite apart from these considerations, family members must also deal with the prospect of both their own death (hopefully, at some point much later in time) as well as losing a loved one. The realization of these facts is usually quite sufficient to produce grief that is capable of not only bringing us to tears, but to our knees as well.

Although the emotional burden that one feels in providing care, love, and support to a dying family member is often inestimable, there are other significant ways in which the members of one's family can be burdened by a decision to put off death (frequently, as Hardwig notes, when there is little chance that the effort will prolong life). An individual in failing health who may not, for one reason or another, be cared for in a hospital setting can seriously strain the physical stamina of even the most devoted family member. Caring for someone who is not in a position to see to that person's most mundane needs is much more than a full-time job; it is not as if you can arrive at 8 in the morning and leave at 4:30 in the afternoon. It is common for those who are providing care in the home to go without sleep for several nights out of a week or, at least, have their own sleep interrupted during the night(s). Sleep deprivation is apt to be common in such circumstances, but as significant and unpleasant as going without sleep can be, there is the demand on one's reserves of physical energy to keep in mind. The fatigue that comes from providing what may well amount to around-the-clock nursing care without a change in shifts can be quite difficult to overcome. What is more, a caregiver is often left too tired to pursue his or her usual interests. This, of course, makes things all the more trying. Without a life of one's own and the time and energy to engage in recreation or hobbies, or even some time just to rest and be alone, one is apt to become even more debilitated. Living under such conditions can quickly grow as old as it is tiresome.

As Hardwig further notes, there is an economic side to death and illness that must be given consideration. It is no secret that medical care, particularly at the level we are accustomed to in this society, is extremely expensive. Given a sufficiently serious illness, it is easy to deplete what liquid assets a person has acquired in preparation for his or her retirement, and when one's own assets have been drained, it becomes tempting to ask that family members make an economic contribution, often in addition to any physical and/or emotional commitment. In many cases, family members are all too eager to help in whatever ways they can manage, including providing money to help a loved one offset medical expenses. What is more, it is easy for loved ones to feel regret or worse, guilt, on realizing that there is little in the way of economic support they can supply. Almost certainly, they are apt to feel guilt when they can provide economic support and decide, for reasons that are on the whole legitimate, to refrain from doing so or continuing to do so. Given the cost of high-quality medical care, it would not be difficult to put one's family on the brink of financial ruin. The exploitation of the family's willingness to provide help can lead to a condition in which assets family members have set aside are in peril. Clearly, Hardwig is quite correct to be concerned about the possible economic injuries that a family can sustain. He writes of one instance (which I will return to later) in which a middle-aged daughter was financially ruined and without a job because her mother wanted to avoid death or at least put it off. Indeed, one of Hardwig's most important points is that a person who is facing the prospect of his or her end ought to be mindful of not just the real possibility of becoming virtually penniless, but also the economic risk that prolonging life may pose for the family.

I intend to examine Hardwig's argument that there exists a duty to die (in the relevant cases). Before doing so, however, there are one or two points to be made in connection with Hardwig's use of the phrase, "duty to die." Strictly speaking, when one claims that a duty to perform an action is present, one means that the action is morally required. This is equivalent to saying, and the point is supported by an axiom of deontic logic, that it is morally impermissible to refrain from performing the action. Hardwig, I believe, may not be using the term "duty" in quite this way. At the beginning of his paper, he writes, "As is appropriate to my attempt to steer clear of theoretical commitments, I will use 'duty', 'obligation', and 'responsibility' interchangeably, in a pretheoretical or preanalytic sense."[3] In a footnote appended to this sentence, he offers the following:

> Given the importance of relationships in my thinking, "responsibility"—rooted as it is in "respond"—would perhaps be the most appropriate word. Nevertheless, I often use "duty" despite its legalistic overtones, because [Richard] Lamm's famous statement has given the expression "duty to die" a certain familiarity. But I intend no implication that there is a law that grounds this duty, nor that someone has a right corresponding to it.[4]

I have little trouble with interchanging the terms "duty" and "obligation"; they are often treated as logically equivalent notions. Having said that, it is likely to be at least misleading, and possibly wrong, to interchange "responsibility" with "duty" and/or "obligation" in as much as these are logically distinct concepts. An example may help to make the point a little sharper. A father of two children acts irresponsibly if he drives without wearing a seatbelt, but it is not immoral for him to refrain from wearing a seatbelt. To assert a duty to wear a seatbelt is tantamount to a claim that it is morally wrong for him to refrain from doing so.

There is a second worry: it is not clear what Governor Lamm had in mind when he used the expression "duty to die". I do not know that it is possible for Hardwig to explain precisely what the governor intended, but if one understands the phrase in its technical or quasitechnical sense, then, as I noted, the claim that there is a duty to die is equivalent to the assertion that it is morally impermissible to refrain from dying. Moreover, if Hardwig (as opposed to Governor Lamm) intends a meaning that falls short of a claim involving moral impermissibility, it is still not clear what he is asserting, for it is not clear what he or anyone means when he or she says that one is not being responsible should one opt to continue living. What is more, a charge of irresponsibility, in the absence of clarity, may involve a lesser failing than the abdication of a duty. If that is correct, Hardwig's claim may turn out to be a little less controversial and, to that extent, a little less interesting. Of course, it is not easy to settle the matter conclusively unless and until one comes clean on what is meant by an allegation of irresponsibility.

There are certain other matters, related to the foregoing discussion, that I would like next to examine. On the assumption that Hardwig intends to argue that there is a duty to die (an assumption that is reasonable, since without it, Hardwig's claim is either unclear or less controversial [and hence, less interesting] or all of the above) as opposed to a responsibility to die, questions can be raised about how such a duty can arise. Hardwig does an impressive job in describing and explaining the myriad ways in which a decision to pursue life or forestall death can adversely impact and undermine the interests of one's immediate family. As he notes and I affirm, there are serious, even crushing, economic, physical, and emotional woes that can significantly interrupt and forever change a family's fortunes. It is woefully commonplace that families have been wrecked by the catastrophic illness of a loved one. This is a fact that should not go unappreciated. However, for all that, I do not think it is possible to ground a duty in such a fact, as austere and commonplace as it may be, and this, I take it, is a plausible reading of his argument. Hardwig writes:

> A serious illness in a family is a misfortune. It is usually nobody's fault; no one is responsible for it. But we face choices about how we will respond to this misfortune. That's where the responsibility comes in and fault can arise. Those of us with families and loved ones always have a duty not to make selfish or self-centered decisions about our lives. We have a responsibility to try to protect the lives of loved ones from serious threats or greatly impoverished quality, certainly an obligation not to make choices that will jeopardize or seriously compromise their futures. Often, it would be wrong to do just what we want or just what is best for ourselves; we should choose in light of what's best for all concerned. That is our duty in sickness as well as in health. It is out of these responsibilities that a duty to die can develop.[5]

However, the fact that another party incurs a burden if I fail to perform some action does not entail that I have a duty to perform the action in question. Consider, for example, the following case: suppose there is a large parcel of property with a river running through the middle of it. There is a bridge over the river, which Smith uses in order to stay dry when he is fishing. Imagine further that I purchase the property and tear down the bridge. Smith will now be burdened (at least inconvenienced); he will either need a new fishing hole, be confined to one side of the river, or get wet. Even so, I have no duty to leave the bridge intact; it is after all my land and my bridge.

Hardwig is not likely to be impressed by this example (and he should not be; I have not said much to put him in harm's way). His most likely reply is that I am not being fair to his case; his argument for a duty to die is predicated on the plausible assumption that additional burdens placed on one's family, assuming a decision to put off death, are apt to be near overwhelming. In the case I have given, Smith is made to suffer an inconvenience, but he is not significantly burdened. I accept this point, but will Hardwig's argument be more convincing if we assume a serious burden, as opposed to, say, a mere inconvenience? I do not believe it will. Consider the following: suppose the same river but it is located on a much larger piece of land, say, 500 million acres (never think small!). I own the land, the river, and a dam that regulates the flow of water downstream. Two thousand miles downstream, we may imagine, is a tribe of people who very much depend on the river for water to drink and to nourish their crops. But imagine I want to create an absolutely enormous lake, the biggest ever (who knows why—just suppose). I shut the water off and soon there will be no downstream to speak of. The tribe living two thousand miles away is going to suffer a serious burden; they will have to find another source of water for nourishment, and their life-style and local economy will undergo significant change. However, I have made no agreement with them to provide water and, as the expression goes, it is my land, my dam, and my river. It seems, furthermore, in the absence of other facts (and there are no other material facts), that I do not act immorally if I prevent, with no specifically intended consequences, the free flow of water to the tribe downstream. Therefore, by an axiom of deontic logic, I have no duty or obligation to do so. Now it is true that I am callous, uncaring, brutishly insensitive, and excessively self-regarding (assuming awareness of the tribe), but I do not act immorally and I am not shirking a duty. As Judith Thomson might have put the point, "What claim do the people downstream have against me that I should provide them with water? It would be gracious of me to do so, indeed it would be the decent thing to do.

But I do not act unjustly (or immorally or avoid any duty) if I fail to provide them water."[6]

Is it possible to sustain so much as an allegation of irresponsibility in the present case? I think there is a sense in which it can be said that I have behaved in a manner that is irresponsible. I have not merely inconvenienced the tribe downstream; they have been made to suffer a burden of some consequence, but it is worse than being made to suffer a burden; it can be said I have acted irresponsibly in that my action may result in the loss of life (assuming I take no steps to determine what consequences to others may follow from my actions). However, there is another sense of the term, and it is one that is hard for me to hear, which (despite its being hard to hear) probably does not apply. There is nothing present in the case I have given that makes me think I am responsible for providing the tribe with water. There is no agreement to do so. The tribe has not, by hypothesis, exchanged anything of value for water. Therefore, it is difficult to cite some fact that entails that I have such a responsibility. At least, I know no such fact.

As far is I can tell, a duty, obligation, or the responsibility to do something (or refrain from doing it) does not, without exception, arise from a consequent burden to others that my action or inaction may bring to others. Duties and obligations (and, I suppose, responsibilities) can, it is certain, be incurred, but obligations and responsibilities are often established through some sort of formal or quasi-formal mechanism. For instance, I can incur a duty through the making of a promise, and there are conventions that govern the making and keeping of promises, which, it is worth noting, have little to do with the distribution of needs and burdens. Or, I can incur an obligation to perform an action where the obligation is grounded in some principle of morality. For example, it has been noted that the principle of utility may be extremely difficult to abide by; some have thought it excessively demanding.[7] Nonetheless, if the principle is correct, then I may be required to perform an action that undermines my interests for the sake of promoting aggregate utility. When this happens, and it may not be rare, I will have incurred a duty to act in a matter that is inimical to my interests. Interestingly (in some cases), aggregate utility may be promoted by bringing one's life to an end. Certainly, Hardwig may be grounding his argument on such local utilitarian considerations, but local utilitarian considerations generate a duty to sacrifice your life to save three or even just two of one's mates in a grenade/foxhole case. Such action is arguably not required; it is above and beyond duty.

Let us suppose that "duty" and "responsibility" are not used interchangeably and that they are treated as logically distinct notions. Is there another way for Hardwig to make his case? Consider the following (informal) argument:

1. One has a general responsibility to protect the overall welfare of one's family.
2. One has a duty to make choices and undertake actions that are consistent with the general responsibility to protect the overall welfare of one's family.
3. In certain cases, the overall welfare of one's family would be compromised, if one chose to put off death.
4. Therefore, in such cases, one has a duty to end one's life expeditiously.

Perhaps this is closer to the argument Hardwig actually has in mind. In any case, the basic idea is simple: responsibilities generate duties.

Although this argument might work a little better, there are problems nevertheless. First, it is not obvious that a (one-sided) duty to die emerges from a responsibility to protect the overall welfare of one's family. The main reason is this: family members have a duty to themselves, if self-regarding duties make sense, to see that their vital interests and general welfare are not undermined. A family member has something to say about the extent to which he or she is going to allow himself or herself to be exploited or otherwise burdened by the pleadings of a doomed relative. Also, although it is true that family members often labor under diminished capacity and, consequently, are not in the best position to protect their interests, if diminished capacity absolves a family member of his or her duty to himself or herself, then the diminished capacity of the patient ought likewise to absolve him or her of certain obligations, e.g., a duty to die (particularly when incapacitation, emotional and/or physical, prevents a person from satisfying the obligation). In addition, the patient's diminished capacity is at least a mitigating circumstance and, quite possibly, it is one that contributes to the notion that his or her action in choosing death can be seen as heroic and, hence, supererogatory. There is, for example, no small amount of fear and anxiety to overcome if one is to let go of life and submit to death. Fighting off contrary emotional impulses to do precisely what is best for one's family may take all the courage one could ever hope to have, but this is exactly how an action comes to be regarded as heroic—it is understood that heroic action is not altogether bereft of fear; far from it, the distinctive element in heroism is that quite in spite of fear the right thing is (somehow) done.

Second, as I said above, the members of one's family have a duty (or, to borrow Hardwig's expression, responsibility) to protect their general welfare and their interests—psychological, financial, and physical. It can be argued that this duty is a primary one, primary in the sense that it is not subject to pre-emption. Still, it is not necessary to pursue such an argument since, I take it, few would maintain that a person's family is required, that is to say bound by duty, to subject themselves to physical, financial, and emotional ruin to extend the life of a loved one. That one's family is exempt from any such requirement has some plausible, but perhaps harsh implications. If a patient's family finds they are overtaxed and are simply not up to providing fully for the needs of a loved one, they can always resort to Medicaid. In many cases, a family would prefer virtually any other alternative. Even so, it is an alternative that exists in both logical and moral space; turning to Medicaid to avoid a significant depletion of their assets is something that they may do. Whether a family has the emotional wherewithal to pursue this option is often an open question, although sometimes it is not; many families would flatly decline to take such measures. If they do decline, they must bear some responsibility for making that choice.

Third, in cases where a family is burdened by the choices and actions of a loved one, such burdens cannot always be foreseen and are nearly always unintended. If the burden is largely unforeseen, then I would be inclined to argue that the patient is absolved of his or her (I am supposing, *prima facie*) duty to die. He or she cannot be expected to pre-empt what he or she does not see. If the burden is foreseen, unintended (almost certainly) and the choice to die is made with the explicit purpose of protecting the general welfare of one's family, then the choice is arguably heroic and, consequently, not obligatory. Consider the following: if my child is in mortal danger and I can save her only if I sacrifice my life for hers, would not doing so involve an act of heroism? If so, there is no duty involved. Moreover, if there is no specific duty (in contrast to an option) to sacrifice my life to save the life of my daughter (and I honestly do not see that there is; at least there is no such duty grounded in any principle of morality known to me), then how can there be a duty to die based on becoming a burden to one's family, if the burden my daughter must bear (should I decline to save her by exchanging my life for hers) exceeds the burden my family endures (should I decline to die)? If the distribution of burdens plays a fairly direct role in producing a duty, as Hardwig seems to contend, there had better be a calculus where the extent to which a party is burdened makes some sort of rational difference in whether a first-person duty emerges.

Hardwig discusses three objections to his claim that there is a duty to die. The most serious of the three, at least in my view, runs as follows: "seriously ill, debilitated or dying people are already bearing the harshest burdens and so it would be wrong to ask them to bear the additional burden of ending their own lives."[8] I want to make some comments about Hardwig's treatment of this objection, particularly inasmuch as the relative distribution of burdens is a topic that has surfaced in the preceding. In the course of discussing this objection, Hardwig has occasion to mention an 87-year-old woman dying of heart disease. He writes:

> An 87 year old woman was dying of congestive heart failure. Her APACHE score predicted that she had less than a 50 percent chance to live for another six months. She was lucid, assertive, and terrified of death. She very much wanted to live and opted for rehospitalization and the most aggressive life-prolonging treatment possible. That treatment successfully prolonged her life (though with increasing debility) for nearly two years. Her 55-year-old daughter was her only remaining family, her caregiver, and the main source of her financial support. The daughter duly cared for her mother. But before her mother died, her illness cost the daughter all of her savings, her home, her job, and her career.[9]

Hardwig goes on to note, contra the objection, that the burdens appear to be unevenly distributed. Which of the following would we judge to be the greater burden: (1) To lose a 50 percent chance of six more months of life at age 87 or (2) to lose all your savings, your home, and your career at age 55? Hardwig claims that (2) is the greater burden. Relative to (1), it would appear he has a point. However, he has stacked the deck. It is not merely that the mother stands to lose a 50 percent chance of 6 more months of life at age 87. The mother is at risk of losing her life (actually, she would have lost two years of life, however debilitated her condition, while overcoming a 50 percent chance of losing her life within 6 months). She must deal with the uncertainty, anxiety, and fear associated with the prospect of no longer living. This is clearly a considerable burden. Unless one is writhing in agony, coming to terms with the end of one's life is usually the most difficult task human beings must face (although, arguably, grieving over the loss of a loved one may be harder). To be sure, what the daughter has endured is not to be under appreciated. Furthermore, there is a plausible argument to the effect that mother and daughter alike would have been left in better circumstances if only the mother could exhibit the courage it would take to face death directly, but she had evidently given in to fear and could not find it within herself to deal squarely with ending her life.

Her fear is an important component in the picture. It is the very thing that prevents her from fulfilling an alleged duty to die. I am moved to wonder, however, whether there is a version of ought implies can that can be invoked in the present case. For example, is an individual who is morbidly fearful of water, but who can swim, required to save a drowning child if the only way to do so is swimming to the child? I am assuming that he or she is not only able to swim, but that he or she can successfully save the child from drowning without risking his or her life. Therefore, physically, he or she quite clearly can accomplish a rescue but psychologically he or she cannot, or so I am supposing. However, fear can render a would-be hero or heroine nearly motionless. In any case, there is a clear sense of "can" (as it occurs in the formulation, "ought implies can") in which it makes sense to say that the older woman cannot come to terms with the fact that her life is nearly over and that, for her daughter's sake, it would be best to succumb. How, then, is she to comply with a duty to end her life?

In the preceding, I have attempted to argue that there is no general duty to die. I have suggested that those who can summon the courage required to face the end and, consequently, to refrain from putting it off are engaging in conduct that is heroic. If this is correct, then there is no duty to die. In addition, I have been concerned to make the point that a family ought to take some responsibility to see to its interests, as difficult as this is apt to be under such very trying circumstances. Ultimately, neither the patient nor his or her family is in an enviable position; although, I must say, if it were a matter of choice (which, of course, it is not), I would opt for the family's position—at least their encounter with nothingness is going to be put off for a time.

Notes and References

[1]John Hardwig (1997) Is there a duty to die? *Hastings Center Report* 27, no. 2: 34–42. The quote is from page 35 of Hardwig's paper. The emphasis is mine.

[2]Hardwig, p. 36.

[3]Hardwig, p. 34.

[4]Hardwig, p. 42.

[5]Hardwig, p. 36.

[6]The reference is, of course, to Judith Jarvis Thomson's (1971) A defense of abortion. *Philosophy and Public Affairs* 1, no. 1 (Fall): 47–66.

[7]For an impressive defense of how demanding morality can be, one can consult Shelly Kagan's (1989) *The Limits of Morality.* Oxford University Press, Oxford. A nicely summarized account of his view is provided by Mary Mothersill's *J. Philosophy,* no. 10 (October): 537–544.

[8]Hardwig, p. 37.

[9]Hardwig, p. 37.

Abstract

Judith Jarvis Thomson's famous violinist analogy to justify a woman's right to abort an unintended pregnancy may not have succeeded in its primary aim, since it can be argued that she did not properly distinguish between killing and not saving someone's life, but it is very useful to use as a starting point in exploring the issue of whether we ever have a duty to die. In this chapter I consider this well-known case from a different perspective, as well as five other cases. In each of the six cases, it might be said that a person has a duty to die. I use these cases to explore our intuitions on this subject and, at the end, I draw some conclusions about whether one could conceivably ever have a duty to die.

I argue that it is extremely difficult to find a case (although not impossible) in which it can be said that we have a duty to die. I think that it can also be said that for the most part, the stronger the obligation to perform an action that will lead to one's death, the less likely it is to be a situation that could be characterized as having a duty to die. Although it is generally true that we do not ever have a duty to die, sometimes we do not have the right to something that may be necessary to sustain our lives.

Do We Ever Have a Duty to Die?

Susan Leigh Anderson

Judith Jarvis Thomson's famous violinist analogy to justify a woman's right to abort an unintended pregnancy may not have succeeded in its primary aim, since it can be argued that she did not properly distinguish between killing and not saving someone's life,[1] but it might be very useful to use as a starting point in exploring the issue of whether we ever have a duty to die. In this paper I shall consider this well-known case from a different perspective, as well as five other cases. In each of the six cases, it might be said that a person has a duty to die. I shall use these cases to explore our intuitions on this subject and, at the end, I will draw some conclusions about whether one could conceivably ever have a duty to die.

I

Let us begin by recalling Thomson's case:

> [Let] me ask you to imagine this. You wake up in the morning and find yourself back to back in bed with an unconscious violinist. A famous unconscious violinist. He has been

> found to have a fatal kidney ailment, and the Society of Music Lovers has canvassed all the available medical records and found that you alone have the right blood type to help. They have therefore kidnapped you, and last night the violinist's circulatory system was plugged into yours, so that your kidneys can be used to extract poisons from his blood as well as your own. The director of the hospital now tells you, "Look, we're sorry the Society of Music Lovers did this to you—we would never have permitted it if we had known. But still, they did it, and the violinist is now plugged into you. To unplug you would be to kill him."[2]

Let us look at this case from the perspective of the violinist, rather than the kidnapped healthy person. Suppose that you are the violinist and that you did not give consent to being hooked up to another person in this way. (Being at death's door, you were unconscious when this occurred.) Now, let us imagine that you have just regained consciousness as a result of having your circulatory system plugged into the system of a healthy person. Let us assume, further, that this other person is a woman and that you awaken to hearing her protest, in the strongest possible language, the way she has been treated. She *demands* to be immediately unhooked from you. What if the doctors, concerned about what will happen to you or fearing a lawsuit, will not unhook the two of you without your consent as well as hers? Are you obligated to agree to be unhooked? If so, does this amount to having a duty to die, since if you are unhooked from her, it appears that you will certainly die?

It seems to me, whatever problems there might be with her claiming that this case is analogous to at least some cases where women desire an abortion, Thomson has established that one does not have the right to the use of another person's body without that person's consent, even if one needs the other's body for life itself. The right to life, whatever it includes, cannot include having that right. Therefore you do not have the right to use the woman's body, since she has not given her consent. You should, therefore, ask to be disconnected from her.

Does it follow that you have a duty to die? No. Let us suppose that, after being disconnected from the woman, you miraculously continue to live. All those involved should be happy about this state of affairs. This seems to show that your obligation is not to *die*, but not to be a party to an arrangement that would involve using another person's body without that person's consent. So this is not a case in which it can be said that one has a duty to die.

II

Let us consider another case. Suppose that you are a 75-year-old woman who is about to die unless you have a heart transplant. After years of waiting, you have finally made it to the top of the heart transplant priority list. You and your family are overjoyed to hear that a suitable heart has been found, just in the nick of time. But, just before you are about to be wheeled into surgery, a woman comes rushing into your room and begs you to let her desperately ill 14-year-old son, who is next in line on the heart recipient list, have the heart, which will be suitable for him as well. She argues that you have had a full life, whereas his life is just beginning, so he should have the heart. With the new heart, he is expected to be able to live a long normal life, whereas you have fewer years left in any case.

Let us further suppose that if you turn down the heart, letting it go to the 14-year-old boy instead, it would result in your name being removed from the list of heart donees, so it would amount to a death sentence for you. (We could, alternatively, suppose that the odds of another suitable heart becoming available before you die are virtually nil.) Should you give the heart to the boy, and if so, does this amount to having a duty to die?

It seems less clear than in Thomson's case that you have an obligation to agree to the request in this case; but it can be argued, on utilitarian grounds, that the right thing to do is to give up your claim to the heart in favor of the boy. Would this mean, however, that you have a duty to die? Once again, this does not seem to be what is required of you, since if you happened to continue to live, everyone would be delighted.

There is, however, something different about this case, compared with the previous one, because of the introduction of the age factor. Daniel Callahan has maintained, given scarce resources, "our collective social obligations to each other" and a proper attitude toward the aging process, that older persons do not have a right to heroic life-sustaining care. He rejects the current attitude that "think[s] of aging as hardly more than another disease, to be fought and rejected" rather than accepting the aging process that "should be in part a time of preparation for death."[3]

> [B]eyond the point of natural life span, government should provide only the means necessary for the relief of suffering, not life-extending technology. By proposing that we use age as a specific criterion for the limitation of life-extending health care, I am challenging one of the most revered norms of contemporary geriatrics: that medical need and not age should be the standard of care.[4]

If we accept Callahan's view, does this mean that older persons who need extraordinary care to sustain life no longer have a right to life? If so, do they then have a duty to die? Once again, I do not think that this would be the correct way to represent the situation. An important distinction needs to be made between *having the right to something that is necessary to sustain life* and *having a right to life*. The elderly person may have lost the former right, if the "something" is a scarce commodity and needed by younger persons, but (s)he has not lost the latter right. This can be shown, as with the first two cases, by pointing out that no one intends the death of the older person. If (s)he somehow manages to continue to exist without the believed to be necessary life-sustaining treatment, everyone would be pleased. Therefore, the obligation is not to die, but to refuse heroic life-sustaining care to which others are more entitled.

III

It might be thought crucial, both to establishing that one has no right to life-sustaining treatment under certain conditions and to its not being described as a duty to die, that those conditions include one's having no prior claim to whatever is needed to sustain one's life. In the cases considered so far, this has been true. The violinist had no prior claim to the use of the woman's body, nor did the 75-year-old woman have a claim to the use of the particular heart that was being donated *before* her name reached the top of the heart recipient list.

Let us now consider a third case, exactly like the second case except that what you need to continue to live is not a new heart, but some scarce equipment to which you are already hooked up. The mother begs you, again on the grounds of your having already lived a full life whereas her child is at an earlier stage of life, to give consent to be unhooked from the equipment so that her son can use it, even though she recognizes that you will certainly die once this life support is removed. Are things different because the equipment had previously been allocated to you and you are now being asked to give it up so that another person, who could make better use of it, might live?

I think the changed circumstances do make a difference. There is a more direct causal relationship between what you are being asked to do and your death, and I think this somewhat weakens the obligation to do it. Many, however, will still feel that you are obligated to help the boy, so you should ask to be disconnected. (I suspect that Callahan, for instance, would.) It is getting closer to being described as a duty to die, because of the more direct causal relationship between what you are asked to do and your death; but I still believe that, since no one desires to bring about your death, it should not be described in this way.

IV

In all three cases that we have considered thus far, it can be maintained that there is not a duty to die, because death is an undesired, and therefore *unintended*, consequence of an action that it can be claimed that you ought to do under the circumstances. At this point it would seem that one is able to drive a logical wedge between the obligation and the death that just happens to result, but is not desired by anyone, because the person needs something that is *external* to him/herself to continue to exist.

In my final three cases, this will not be true. Each case will involve others having need of (parts of) your body, so that if you have an obligation to satisfy that need, it must be thought of as involving the necessity of your death. Thus, it would be appropriate to say that if you have the obligation, it is a duty to die. I shall introduce these final cases in an order arranged according to (arguably) increasingly stronger intuitions that one should sacrifice one's own life under the given circumstances.

Here is my fourth case: You are a healthy, but quite ordinary single person who has no children. A very important person, on whom many others depend, is desperately in need of a new heart and it has been determined that only your heart could save him. (It is the only heart in existence that could provide an acceptable tissue-type match.) Should you agree to give up your life to save him?

My intuition in this case is similar to Thomson's about her violinist case:

> Is it morally incumbent on you to accede to this situation? No doubt it would be very nice of you if you did, a great kindness. But do you *have* to accede to it?... I imagine you would regard this as outrageous.[5]

Perhaps it is a bit too strong to say that to have to feel obligated to accede to the request is "outrageous," but it is certainly above and beyond the call of duty. Of positive responses to requests like these, however, heroes/heroines are made. (Even more heroic would be to volunteer before one is asked.) Still, it would be hard to maintain, unless one is a strict utilitarian, that one has a *duty* to die in this case. (This example, and others like it, show that a strict utilitarian theory may be unacceptable.)

What if we increase the number of people who could survive only if you sacrifice your life? In my fifth case, let us suppose that you alone have the right tissue type to supply organs to five important people who will die without these organs. They happen to need different organs so that your body alone could satisfy all the requests. Let us suppose, further, that your dying would have little negative impact on others. No matter how important the five people are, and no matter what the loss to rest of the world if they do not survive, intuitively it does not seem that you have a duty to die (again, unless you are a strict utilitarian, which seems indefensible), but perhaps it becomes a little more plausible to ask that you consider making the sacrifice.

V

In my last case, I will take my best shot at coming up with a case where it might reasonably be thought that one has a duty to die. Suppose that you have contracted a new, highly contagious disease that has the potential for destroying humanity. You have the most advanced case of this disease, which is quickly spreading throughout the world, so examining your body will yield the most information about this disease. The only hope for heading off a global disaster is for scientists to autopsy your body as quickly as possible (let us suppose that it is either too dangerous to study your body while you are alive or that they can only get the information they need in this way) to see if they can figure out how this disease can be cured, or at least brought under control. The Center for Disease Control has begged you to allow them to euthanize you as quickly as possible, arguing that you are condemned to die in any case, but through your hastened death you could save humanity. Do you have a duty to die?

I have tried to concoct this case so that anyone who has at least a tiny bit of a utilitarian component to his or her ideal ethical theory[6] will feel a strong pull in the direction of saying yes to this question. More particularly, it is also important to this case that by giving up one's life, one could prevent great harm from being done to others, rather than just bring about more good consequences. As W. D. Ross argued in *The Right and the Good,* the obligation to cause the least harm through one's actions is stronger than the obligation to bring about the most good.

What do I think about this case? I personally would feel that I had a duty to die under these extraordinary circumstances, but I could understand it if you happened to believe that one's life is a gift from God and that it is always wrong to end it prematurely. (Perhaps you also feel that God is aware of, and so most likely intends, what is happening. You may believe that it is part of God's plan that the human race should cease to exist.) I on the other hand, and I think many others as well, believe that "[m]oral goodness has something to do with the amelioration of suffering... and the promotion of human flourishing."[7] Therefore, it does seem to me that under these extraordinary circumstances, you do have a duty to die.

VI

What can we conclude from examining these cases? Generally speaking, it is extremely difficult to find a case in which it could be said that, we have a duty to die. I think it can also be said that, for the most part, the stronger the obligation to perform an action that will lead to one's death, the less likely it is to be a situation that would be characterized as having a duty to die. In Thomson's case, the one in which there is the clearest obligation to do the action that is likely to bring about your death, you ought to ask to be unhooked from the woman's body, but we certainly do not think of this as having a *duty to die*. If you die, as is anticipated, you just *happen to die* from being unhooked from a body that it was not right for you to be hooked up to in the first place. In the next two cases I presented, it can be argued that you have an obligation to refuse the heart transplant or ask to be removed from the equipment that is believed to be sustaining your life, but the obligation becomes more questionable. There is an increasing feeling that what is being asked of you is getting closer to a duty to die, but it does not quite reach that point.

As we move to cases in which it seems more appropriate to say that *if* you have an obligation to do the action, it could be described as having a duty to die (such as a case where others desperately need your organs), we are less likely to say that you are *obligated* to do the action. We would, instead, think of it as an heroic act, an action that is above and beyond the call of duty. Throughout these examples, it seems that Thomson's main point that others do not have a right to use your body without your consent holds up. If others do not have a right to use your body, then you are not *obligated* to sacrifice yourself by letting your body be used to save others. It might be wonderful if you were willing to do so, but you do not *have* to.

The only circumstances under which it would appear to most ethicists, I believe, that one might have an obligation to die would be circumstances that we are extremely unlikely to face in real life. These circumstances would include the near certainty of great harm being caused to a large number of persons if one remains alive, in which only one's death could save these people (and it helps that one is likely to die soon in any case). I have attempted to describe a case that would include these circumstances. Short of this extremely unlikely set of circumstances, I would say that we never have a duty to die, but sometimes we do not have the right to something that may be necessary to sustain our lives.[8]

Notes and References

[1]It can be maintained that Thomson has shown that one may not be obligated to save the violinists life, but this does not entail that a pregnant woman is justified in taking the life of the fetus, which is a very different matter. *See* Baruch Brody (1975) *Abortion and the Sanctity of Human Life: A Philosophical View.* M.I.T. Press, Cambridge, MA and Francis J. Beckwith (1993) *Politically Correct Death: Answering the Arguments for Abortion Rights.* Baker Book House, Grand Rapids, MI.

[2]Judith Jarvis Thomson (1984) A defense of abortion, in *The Problem of Abortion,* 2nd ed. Joel Feinberg, ed. Wadsworth Publishing Company, Belmont, CA, p. 174.

[3]Daniel Callahan (1996) Aging and the ends of medicine, in *Biomedical Ethics,* 4th ed. Thomas A. Mappes and David De Grazia, eds. McGraw-Hill, New York, p. 579.

[4]Callahan, *op. cit.*, p. 581.

[5]Thomson, *op. cit.,* p. 174, 175.

[6]Although a strict utilitarian theory is problematic, I think completely eliminating any considerations of consequences is equally problematic.

[7]Louis P. Pojman (1996) The case for moral objectivism, in *Do the Right Thing,* Francis J. Beckwith, ed. Jones and Bartlett, Sudbury, MA p. 17

[8]None of what I have said in this chapter should be construed as showing that one never has a *right* to die, which is an entirely different matter from whether one has a *duty* to die.

Abstract

Earlier literature on the duty to die is meager but emphatically promotes such a duty. According to preceding chapters, this duty rests on two foundational principles: the beneficence principle of not burdening one's loved ones and the justice principle of not consuming more than one's share of scarce resources. Supporters of this duty argue that its implementation can be made at the level of public policy and/or at the family level. They also argue that attitudes concerning the duty to die can be made more acceptable through socialization, and that although this duty exists, it has been thus far unrecognized. Following their arguments closely, I explore the reasoning that supports the duty to die, whether the duty should be limited to the elderly or to the ill in general, the implications of recognizing duties not universally recognized, the notion of "value" that underlies the concept of "scarce resources," and the consequences of implementing this duty. I raise questions about why the foundational principles of beneficence and justice do not entail other duties for persons in affluent societies, and whether this duty does not leave us with an idiosyncratic view of "duty" itself. I come to the conclusion that with all of its novelty, the ramifications of recognizing such a duty leave us in a strange moral situation indeed.

Grandma, the GNP, and the Duty to Die

Judith Lee Kissell

The concept of a duty to die is a relatively new one for medical ethics. It raises questions about whether, given the current state of health care finance, demographic projections of the elderly, and our increasing technological ability to keep people alive and functioning, we may not at some point be obliged to simply die. Do we have an ethical obligation, like the Eskimos who, when they become old and feeble, simply lag behind and float off on a piece of ice? A few ethicists have lucidly and compellingly advocated just such a duty.[1] I will depend to a great extent on these pacesetters of moral thinking, granting that their positions are well and persuasively argued, and I will employ their reasoning. I will further allow, second, that the duty of which my predecessors speak in these excellent papers should be acknowledged on a par with other well-recognized duties (such as preventing harm or helping one's neighbor in trouble). That is, we should not minimize this obligation or relegate it to secondary importance, but recognize it as an authentic, full-fledged, moral imperative. I will assume that although various historical contingencies have so far obscured it from our moral vision, the duty to die is a *bona fide* obligation. Third, I will concede, just as these innovative thinkers do,[2] that this duty should be implemented. My task will be to explore how and at what level we should put it into practice, how this concept fits into the traditional concept of duties, how concrete decisions about operationalizing it should be made, and what advantages of such implementation will accrue to us as a society.

However, before I undertake this job, I must locate and clarify the principles that support the positions of my predecessors and on which implementation must be based. The duty to die, as it has been so far articulated, is grounded in and rests on, two other principles. First, the beneficence principle: we are obligated to limit the manner and degree to which we burden others. A second and subsidiary justice principle asserts that we must not absorb an inordinate amount of health care resources; by limiting our own consumption, we might make available to others a fair share of health care. As jointly sufficient conditions, both the beneficence and the justice factors must be present for the duty to exist. In other words, a responsible father has an obligation to end his life, or his life-extending treatment, well before he has ruined his children's marriages and careers and usurped their children's (his grandchildren's) college funds. This obligation exists when both conditions are fulfilled: he poses an emotional and financial hardship on his family and he absorbs an inordinate (unjust) amount of their resources and/or of health care resources in general.

Some might protest that if minimization of burden and fair allocation of resources are behind the duty to die, why do these principles apply then only to healthcare and why not to food, water, land, and so forth. A more problematic objection would be this: if the duty to die stems from the overconsumption of health care resources, does this principle not obligate us as Americans, no matter what our age or condition, to consume less? Should we all be more frugal in our use of health care so that less prosperous persons, particularly children either from underdeveloped countries or from sectors of our own country—Native Americans, for instance—might have their fair share? We might more easily fulfill this latter obligation by simply forgoing such procedures as face-lifts, liposuction, and fertility treatments though, rather than by committing suicide.

Moreover, one could object that virtually all the burdens of remaining alive too long would be much mitigated by universal health coverage combined with more sensible attitudes regarding futile treatment and care for the dying. For instance, George Soros' "Dying in America" project advocates more compassion toward, and palliative care of, seriously ill persons. Of course, every dollar spent on end-of-life care, even the universal health coverage dollar, still is a dollar denied to furthering such projects as genetic engineering and therapy that would, after all, deliver us from many of the burdens currently imposed on us by the ill, but these are insignificant points, and we must resist being distracted from weightier issues.

The most critical question for the duty to die concerns the practicalities of its implementation. Let us examine the difficulty here. Any reasonable bystander recognizes the difference between the moral obligation to save a child drowning in a foot-deep pool of water and that of, say, diving into the freezing water to save the Titanic passengers. General duties of helping one's neighbor or of preventing harm are mitigated, if not abrogated, when fulfilling these obligations would result in disproportionate inconvenience or risk for the would-be rescuer.[3] In other words, the concept of "duty," though related to beneficence, traditionally has been thought to be regularly associated with self-interest. However, if one were to have a duty to sacrifice one's most basic interest one's life "duty" would become radically disassociated from self-interest; in fact, the very concept of self-interest would lose its meaning.

Given this disengagement from its traditional theoretical roots, the duty to die then seems radically dissimilar, rather than closely related, to other commonly acknowledged duties. Additional understandings about this notion may fall by the wayside as well. We may be bound in ways that we do not yet suspect, but that historical circumstances or at least someone's interpretations of historical circumstances may impose on us. Therefore we may ask by whose definition is health care a scarce resource? Might this definition not be simply a manipulation of the entrepreneurial health care market that reflects the interests of Wall Street a kind of society-wide downsizing or trimming of the fat. Surely health care is not scarce in the sense that water is scarce in Death Valley and might be squandered on some useless project. Does its "scarcity" not rather hinge on what is considered valuable, and upon who makes that determination?

We do not, for instance, hear many complaints about the use of scarce public resources by couples who consume expensive fertility research and treatment that they could themselves afford, or about the amount of research dollars that go into treatment for sexual impotence instead of into the less lucrative market for treatment of hepatitis C. Newspapers are filled with advertisements for the latest pharmaceutical products or for the newest techniques in cosmetic surgery aimed at middle-class consumers who can afford them. Why are these resources not considered scarce, and why is their use not considered waste? Are the elements of our lives that are thought to have value exclusively limited to those that contribute to the GNP, as numerous current public policy debates claim? Therefore, for instance, just as the contributions of stay-at-home wives and mothers are considered by Social Security policy not to be valuable, since they fail to increase the GNP, so components of health care that promote societal solidarity or that relieve the anxieties of worried parents are likewise considered to lack worth. Who in our society, after all, decides that GNP determines value?

That the duty to die has not until this point in history been recognized by Western society raises the question of what other duties we have been too myopic to detect. Might there be additional obligations that would be inimical to our life and limb, but that other members of society—who better understand these matters—feel morally obligated to help us observe? One advocate[4] of the duty-to-die leaves open the question of whether those of us who are intellectually competent might have an obligation to help the incompetent fulfill their duty, but what about those who are mentally competent while remaining morally incompetent? Are we who are morally more sensitive and proficient obliged to help these others fulfill their duties?

Some might, for example, believe that members of society are bound to donate organs whether they want to or not. We might even be forced to donate a nonpaired organ, such as a liver, to someone who contributes more to the society than do we, or some persons (perhaps those who are unemployed, uninsured, a burden on their families, and leeches on the society) should be helped (encouraged or coerced) to accept a physician's assistance with suicide. In view of overpopulation in many Third World countries, should parents of two children be obliged to submit to sterilization? Conversely, in view of declining birth rates in many Western countries, might such a view justify a husband's raping his wife under the auspices of a duty to bear children?

Traditional understanding of this concept has entailed universizability, that is, obligations apply equally to all relevantly similar persons and circumstances. However, the conditions that ground the duty to die might violate this principle. For someone with sufficient money to buy any desired health care, who has no family who cares about him or her and who in fact, provides employment for a paid care giver, seems absolved of the duty to die, for he or she fails to satisfy the beneficence and justice conditions that ground this obligation. Once the concept of duty, as instantiated in the duty to die, is loosed from its traditional moral moorings—its relationship to self-interest, its basis in rationality rather than historical happenstance, its association to universizability—it is difficult to think where our moral inspirations and innovations will take us. But I digress.

Let us return to the issue of implementation. The existing literature offers at least two possibilities. One school of thought[5] advocates that the data should be exercised at the level of public policy.

The public policy notion would require a procedure that is objective in the best and most moral sense: decisions should be made by third parties who are, presumably, impartial, detached, disinterested, and, I imagine, morally competent. (Although one of my predecessors suggests that decision makers operate from behind Rawls' "veil of ignorance," it is difficult to imagine how that can be arranged in the real world.) Another school of thought suggests that implementation take place within the family. The family version hints that, when faced with making life-and-death choices, members can make caring, loving decisions that will presumably reflect everyone's best interests but that will in no way result in intergenerational strife, selfishness, mistrust, or fear. (No suggestion is forthcoming about what to do with people who have no families. Dysfunctional families are apparently on their own.)

Public policy implementation would be difficult to envision in the currently unsettled health care scene. Granted the impracticality of putting in place the "veil of ignorance," problems regarding who would formulate such a policy seem problematic. No doubt we would be able to find sufficiently objective, impartial, detached, and disinterested persons to do so—perhaps experts in the field, such as physicians, ethicists, clergy persons, economists, psychiatrists—but to determine how to accomplish this task lies beyond my capability. Should these experts be men or women, young or old, rich or poor, black or white, First World or Third World, elected or appointed? Conceivably the project would run into obstacles more serious than those that confronted the health care rationing scheme in Oregon, for in the duty to die case, end-of-life health care would likely involve respectable taxpayers and voters rather than merely the indigent, largely nonvoting population characteristically served by Medicaid.

I avoid these political problems, turning instead to the still difficult, but easier task of appointing these arduous choices to the family, to whom, according to one ethicist, such decisions more properly belong.[8] He suggests a family meeting(s). Since the duty to die forcefully exemplifies the fundamental principle of not burdening one's loved ones, it becomes essential to instill this sense of duty at an early age, along with other moral traits, such as caring for one another, honesty, love, truth-telling, and so forth. I am encouraged in this view by another author who, in conjunction with advocating the duty to die, advocates direct killing of the elderly:

> Nevertheless, whether death in old age is feared or welcomed is very much a product of social beliefs and expectations, and these not only undergo spontaneous transformations but can be quite readily engineered. Transformations in social practices in earlier historical periods make it evident that beliefs about whether there is such a thing as a time to die can change...

Surely, her arguments are compelling that the obligation to die when one becomes a burden will more likely flourish if planted early. I will follow her logic to discover the implications that flow from it.

After all, the example of the son who sacrifices his marriage, his career and his children's college fund to care for his father, learned the importance of self-sacrifice while still a child. Moreover, we cannot expect one's elderly parents, for example, to understand and embrace the duty to die just when it becomes convenient for the rest of the family. It seems reasonable to initiate such get-togethers early on, perhaps when the parents have completed their childbearing, and their youngest has reached the age of reason, or when the time comes to say goodbye to a beloved grandparent, who, having exhausted his or her allotted resources, bravely prepares to leave this world.

Some uncertainty exists among my predecessors about whether this duty exists owing to age or to physical disability, although the age factor is strongly favored. In either case, guidance by wise parents is required so that children can be led to making an informed, well-considered, and fair decision about this crucial question (although whether the decision should be made on a democratic basis remains a problem). A clear conflict is apparent if grandparents are allowed to participate in the discussion. On the one hand, grandparents may wield a disproportionate influence over their own children, not to mention their grandchildren. Further difficulties can be anticipated if only one set of grandparents is still alive, or if one set is wealthier, more pleasant, or in better health than the other set.

Other difficulties remain. What about the family of persons like Claude Pepper, who served actively in the House of Representatives, or artist Pablo Picasso, both of whom lived and worked into their 90s? Should the contributions to the society, or to their families, of these extraordinary older persons be taken into account to excuse them from their duty? On the more personal level, should the grandparent who makes a sizeable investment in the children's college fund be let off the hook? What if he or she, while contributing to the family coffers, still requires hours of physical care and significant emotional attention? Which counts more? Should the elderly simply be denied care, in which case they may linger on in pain and disability and actually cause more problems to the rest of the family than if they were treated? Should they be helped to commit suicide, or, as some seem to favor, should they simply be directly killed?[10] Further difficulties arise if the family member resists suicide at the last moment, botches the attempt, or is no longer competent, and then the decision must be made about who kills him or her, or at least, who contacts the family physician.

The second reason that the duty should be inculcated at an early age arises from further problems in the duty to die literature. If the family decides in favor of disability rather than age as the basis of the duty to die, children can be led to realize early the wisdom of terminating lifesaving or life-preserving treatment of disabled siblings. Such self-sacrifice from a young member of their own household would provide an invaluable lifelong example of love, devotion, and true family values. Even very young children can understand that medical costs can quickly absorb money that could otherwise be put aside for college education or a family vacation, for example.

The disability criterion holds other unique difficulties. Decisions must also be made about whether chronic illness that eats up money, time, and attention over the long haul should override acute illness or crisis-type situations, such as accidents. Should attempts to rescue the victim of a serious automobile accident, for instance, be allowed to eat up family and social resources, whether financial or emotional? Again, should family resources be spent on the child who falls into taking drugs? If the answer is yes, should that child get one chance at rehabilitation? Two? Three? What about the anorexic child who refuses to eat sensibly or the one who, by eating too much or by failing to exercise, brings problems on himself or herself? Should children be held strictly accountable for the emotional problems, some perhaps that stem from a misunderstanding of this very duty, but the cost of which burdens the rest of the family? Making these difficult economic, health, and family-life choices, which has traditionally been part of the responsibility of parenthood, can, under this model, be a part of shared family decisionmaking, drawing the family together through mutual goals, in loving harmony and trust.

What about the middle generation—the parent who smokes or drinks excessively? Should the parent who absorbs a large amount of financial and emotional resources, but who still holds down a job, be exonerated from the duty to die? Is this parent excused from his or her duty to die until the children reach an age at which they can take up the burden of caring for the family? Is physical abuse of children a "burden" to the family? Should the duty to die be postponed until working parents are vested in their retirement plan? Should he or she be allowed to live as long as his or her retirement or disability check comes in?

Finally, the increasing ability of medical science to project the health futures of family members stemming from genetic technology provides both problems and opportunities. First the problems: the family must decide, for example, about whether daughters should be tested for breast cancer, or sons and daughters for Parkinson's disease, and at what age? If the family insists on testing to facilitate their crucial duty to die decisions, what psychological help will be given to the unfortunate children or other family members who must now anticipate an unpleasant, perhaps tortuous death? Does the cost of the psychological assistance count against their allocation of health care expenses? Further problems occur since genetic testing may indicate a mere tendency to a disease, or because such testing lacks the capacity to reveal at what age the disease will appear.

An even greater fortuity lurks behind the possibility of genetic testing, providing considerations of yet greater import. The possibilities offered by genetic technologies hold promise of uncovering yet more of the thus far undetected duties mentioned earlier. Such technology boasts the consummate potential for observing the critical moral values that are enshrined within the duty to die. The beneficence principle of not burdening our loved ones and the justice principle of not devouring an inordinate amount of health care resources can, after all, be best adhered to by observing a more pre eminent duty: the duty not to be born.

Notes and References

[1]*See*, for example, Margaret P. Battin (1987) Age distribution and the just distribution of health care: is there a duty to die? *Ethics,* January,317–30; John Hardwig, (1997) Is there a duty to die? *The Hasting Center Report,* March–April; and Richard D. Lamm (1997) Death: right or duty? *Cambridge Q. Healthcare Ethics* June: 111,112.

[2]Governor Lamb hints that at some level of public policy, the economic problems caused by people living too long must be addressed either at the legislative or administrative level. Hardwig suggests that a family sit down together to decide the details of the duty to die in that particular family.

[3]*See,* for instance, Gregory Mellema (1991) *Supererogation, Obligation, and Offence.* State University of New York Press, Albany, N.Y. Joel Feinberg (1984) *Harm to Others.* Oxford University Press, New York, especially Chapter 4.

[4]Hardwig, Is there a duty to die?

[5]*See* Battin and Lamb.

[6]Rawls' "veil of ignorance" shields persons who would make decisions about justice from knowing the circumstances of their lives, e.g. whether they would be healthy or have a chronic disease, whether they would die of cancer or be killed in an automobile accident.

[7]*See* Hardwig.

[8]Hardwig says, "The impact of my decisions upon my family and loved ones is...the most plausible and likeliest basis of a duty to die. 'Society,' after all, is only very marginally affected by how I live or by whether i live or die)" (36).

[9]Battin, Age rationing and the just distribution of health care: is there a duty to die? 335.

[10]*See* Battin's, Age distribution and the just distribution of health care: is there a duty to die?

Abstract

John Hardwig's article, "Is There a Duty to Die?" places the family in a central role in the decision of whether or not a family member has a duty to die. His main criteria for whether such a duty prevails is, briefly, whether we will be a burden on our loved ones. Ironically, it is his insistence on making this a family decision that gives rise to serious problems. The difficulties arise when we consider how a scenario might play out under Hardwig's proposal.

Using one of Hardwig's own examples, this chapter explores the scenario and argues for two conclusions. One is that there are serious, although not fatal, problems with his employment of altruism in the duty to die decision-making process. The other, which is more serious, questions whether we might consider a family that decided one of its members had a duty to die morally blameworthy. Even if we did not, we would still encounter problems when the family informed their loved one that he or she had a duty to die. Is it proper for a family member to ask another to die, or is such a duty more akin to a debt of gratitude, which one might be owed, but it would be improper to demand the gratitude?

In the final analysis, this chapter concludes that the practical application of Hardwig's conception of a duty to die has problems that raise the question concerning whether his proposal is feasible at all.

Dying for Others: Family, Altruism, and a Duty to Die

Ryan Spellecy

I would like to examine John Hardwig's notion in his article, "Is There a Duty to Die?" that having a duty to die is largely dependent on whether or not we are a burden on our family, and by doing so, I hope to clarify two problems with his proposal. The first arises when a family is asked to participate in the deliberations concerning whether one of their own has a duty to die. The other problem stems from the modified form of altruism employed by Hardwig in these deliberations. The first problem concerns roughly what it means to be a family.1 The second problem concerns Hardwig's procedure for determining whether or not one has a duty to die.

Now, let me be clear. I am not disputing whether there can be a duty to die. Certainly, in some circumstances, families ask great sacrifices of their members, and rightly so. There could even be a duty for a family member to die in circumstances like those. Rather, what I intend to do is explore some problems with Hardwig's approach to the issue.

Family

Let us then consider a case in which the family would be severely adversely affected (financially, emotionally, and so forth) by a loved one's illness. For parity, let us employ an example from Hardwig himself.

> An 87-year-old woman was dying of congestive heart failure. Her APACHE score predicted that she had less than a 50 percent chance to live for another six months. She was lucid, assertive, and terrified of death. She very much wanted to live and kept opting for rehospitalization and the most aggressive life-prolonging treatment possible. That treatment successfully prolonged her life (though with increasing debility) for nearly two years. Her 55-year-old daughter was her only remaining family, her caregiver, and the main source of her financial support. The daughter duly cared for her mother. But before her mother died, her illness had cost the daughter all of her savings, her home, her job, and her career.[2]

Hardwig thinks that cases like these are the ones in which a duty to die may arise. I will use this as our example, with a few modifications as we proceed, which will not change the features in a manner unfair to Hardwig. What is most appealing about this example when examining Hardwig's proposal is that the mother "very much wanted to live." If the mother had wanted to die anyway, or if the daughter had decided that her mother was not such a great burden, the case would not be interesting for our purposes. What the case demonstrates is that Hardwig thinks a duty to die may prevail in cases where death is contrary to the wishes of the patient—where the patient very much wants to live.

Let us call the 87-year-old woman Sarah and her daughter Anita. Anita is a successful lawyer with a fast-moving, public-interest group. Her career is such that she would have to quit her job, thereby destroying her career, in order to create the time necessary to care for her dying mother.

The question we may now turn to is whether, after the open and frank conversation prescribed by Hardwig, Sarah will be too much of a burden on Anita, her only remaining family. Let us say that they have had the numerous agonizing and honest discussions recommended by Hardwig,[3] and have concluded that Sarah's illness is such that it would constitute an unbearable hardship for the family, and that Sarah does indeed have a duty to die. How should we respond to such a disclosure? I contend we would be inclined (and rightly so) to at least strongly disapprove of such an action by a family and that this would result from such a family's failure to "live up to" an important element of what it means to be a family.

We would be inclined to condemn such actions, because not only to fail to care for a loved one in such an instance, but to find that after open and frank discussion, she has a duty to die rather than to be a burden is to fail to meet an important element of family. Now, it is important to pay close attention to the distinction just made. What is at issue in our situation is not merely that the family cannot care for its loved one, or that its members are unwilling to try, or unwilling to exhaust all possible avenues of aid. These too would be blameworthy acts, but what is at issue here is much more serious. This is a family that has openly discussed the burdens involved and has decided that one of their family members is too burdensome and has a duty to die.

Dan Callahan, in his letter to the editor of the *Hastings Center Report* regarding Hardwig's article, illustrates this point nicely when he states that, "I believe it trivializes the relationship of family members to each other to act as if their mutual obligations to each other are to be judged by some benefit-burden calculus. Hardwig seems to be saying in effect: 'for better or worse, in sickness and in health—well, sort of, it all depends.'"[4]

Even if you do not share my and I believe Callahan's intuition, you might agree that even if a family could arrive at the above conclusion that Sarah has a duty to die, and not be held morally blameworthy, they still should not or could not express their conclusion. That is, even if a family could determine together that one of their own had a duty to die, it might be wrong to ask one to perform such a duty. In a sense, a duty to die would be much like an obligation of gratitude.[5] I may owe you a debt of gratitude for a favor, but it would be wrong for you to demand that gratitude. Likewise, it would be wrong for family members to inform (and thereby demand) their loved one that he or she has a duty to die. Of course, there are scenarios in which it is indeed proper to ask one to die. A soldier might be asked to die for his or her country. However, this scenario must play out in a different manner than Hardwig intends.

One might have to decide by one's self that one had a duty to die and, "approach your loved ones only after you have made up your mind to say good-bye to them."[6] Hardwig does foresee a similar objection. He counters by saying, "I believe in family decisions. Important decisions for those whose lives are interwoven should be made together, in a family discussion."[7] However, as Hardwig envisions the objection, one decides in isolation in order to spare one's loved ones the pain and grief. That it may be wrong in the sense set out above is another, distinct issue. Thus, the problem remains.

Altruism

With the first objection considered, we may now move on to Hardwig's problematic employment of altruism. For this discussion, let us return to Sarah and Anita. When it comes time for Sarah to decide whether or not she has a duty to die, under Hardwig's proposal she will have to employ a modified form of altruism. I say modified as opposed to pure altruism because some of Hardwig's criteria for whether or not we have a duty to die are clearly not altruistic. Hardwig's criteria can be divided into two categories: those that concern whether you have led a full life and would not be losing much by dying sooner (2 and 3) and those that concern how heavy a burden you will be on your loved ones (1, 4–9). Nonetheless, when one notes that seven of the nine criteria for a duty to die are other-regarding, and reads passages such as, "[T]he genuine love, closeness, and supportiveness of family members is a major source of this duty (to die): we could not be such a burden if they did not care for us,"[8] it is clear that Hardwig's proposal is highly altruistic.

If phrases like "modified altruism" or "highly altruistic" are con-fusing, I apologize. It is not entirely clear what is driving Hardwig's proposal. It may be helpful then to clarify the issue with the following reminder; Hardwig thinks a duty to die may arise even when the patient does not want to die, but desperately wants to live.

Now, I do not wish to suggest that altruism (pure or as employed by Hardwig) is a problematic moral theory or that human beings are unable to act in genuinely altruistic ways. What I do wish to suggest is that altruism as employed by Hardwig and in this context is problematic. In fact, we will have to modify our story once more if Hardwig's proposal is even to make sense (but that will become clear later).

Altruism could theoretically guide my own actions if I were in Sarah's situation. Using altruism as a guiding moral principle, I might come to the conclusion that I ought to act so that I produce the best possible situation for others, with specific emphasis on my loved ones, since they will be the most affected group in my moral considerations. Thus, I could bring the matter up, engage in open and frank conversation with them, and in the final analysis choose that option that would benefit them most. Up to this point, Hardwig's position seems sound.

The problem arises when we consider what my family members are to do or say in this open and frank discussion. Should they too deliberate and act altruistically? I assume they must, since the alternative, that only Sarah is to deliberate and act altruistically, seems bizarre and capricious. It would be strange indeed if one need only deliberate and act altruistically when one will be a great burden on one's family, especially since those like

Sarah are the people already most burdened and most directly affected by the illness. Let us consider this.

> Sarah: "Well, I hate to ruin dinner, but Dr. Miller called today with my test results and I want to discuss this with you, Anita, since you are very dear to me and will obviously be the most directly affected by all of this, except for perhaps myself. After all, you are all the family I have left. I guess I just want to be sure that I am not too much of a burden."

> Anita: "How could you even think that? Too much of a burden? Nonsense. You know I will do all I can. We *will* get through this."

It appears that Anita will have to quit her job (and abandon her career for that matter), and care for her dying mother. After all, this would be her altruistic conclusion. In such a case then, no one would have a duty to die, and no one could be "too much of a burden" on his or her family. Hardwig's duty to die simply could not arise. In families like Sarah and Anita, they could not decide that Sarah has a duty to die.[9]

Now, let me be clear that my objection is not based on what Sarah's family members would do. Although this is an important question, of course, it is not the question at hand. My objection rather is based on what, by Hardwig's account, her family ought to do.

One might attempt to salvage Hardwig's position by replying that Sarah should deliberate alone and consult Anita only to keep her apprised of the situation, but not include her in the actual decision making. Instead, the scenario might run as follows:

> Sarah: "Well, I hate to ruin dinner, but Dr. Miller called today with my test results and I have decided that I cannot be such an overwhelming burden on you, my only remaining family, and so I have decided to take steps to end my life."

This might alleviate the issue of how Anita is to act (altruistically), since her actions would be irrelevant to the decision and thus remove the apparent standstill in the decision-making process. However, such an approach would have its own, distinct problems for Hardwig. Although it might remove the problem of Sarah's family's inability to decide on an altruistic basis that she has a duty to die owing to the great burdens, Sarah would not be acting altruistically if she were to act as such. After all, who would know better what is best for Anita than Anita herself? Should Sarah vaguely and slightly deceptively elicit Anita's opinions, and then scurry away to make her decision in solitude? Certainly not.

Recall a counterobjection posed by Hardwig that has previously been mentioned, with a more complete citation.

> Some may object that it would be wrong to put a loved one in a position of having to say, in effect, "You should end your life because caring for you is too hard on me and the rest of the family." Not only will it be almost impossible to say something like that to someone you love, it will carry with it a heavy load of guilt. On this view, you should decide by yourself whether you have a duty to die and approach your loved ones only after you have made up your mind to say good-bye to them. Your family could then try to change your mind, but the tremendous weight of moral decision would be lifted from their shoulders. Perhaps so. But I believe in family decisions. Important decisions for those whose lives are interwoven should be made together, in a family discussion.[10]

Granted, Hardwig rejects this scenario for different reasons. He considers the situation not because of the family's inability to act altruistically in this situation, but because he altruistically does not wish to impose the hardship of deciding whether a loved one has a duty to die on an already burdened family. Now, I have suggested that Sarah would not be acting altruistically if she were consulting her family, and Hardwig suggests that a motivation for not consulting one's loved ones would be out of concern for their welfare above one's own (altruism). Although this point could be argued further, it does not seem necessary since there is an even greater problem (according to Hardwig's proposal) if Sarah does not consult her family.

Hardwig appears to see this problem when he rejects the above approach, because he believes in "family decisions." Deciding alone, or deciding without open and frank discussion, would be to commit Hardwig's individualistic fallacy,[11] which he surely cannot recommend.

Nonetheless, Hardwig's proposal might still be salvaged. If Anita were not Sarah's only family, but Sarah also had a husband Jack and a son George, the proposal could work. How could our new, larger family avoid the standstill without committing the individualistic fallacy?

The scenario would run exactly as above, except with two more characters. What is significant is that these two extra family members change the structure of the discussion in a manner that enables it to take place as Hardwig describes (at least as far as this section is concerned. The addition of any number of family members would not help Hardwig with the objections contained in the previous section).[12] For example, Anita might now altruistically deliberate taking into account not only Sarah's interests, but Jack's and George's as well. In this manner, the family might arrive at the conclusion that Sarah has a duty to die.

A Duty to Die?

Given the above considerations, it is possible to make consistent sense of Hardwig's position. If the terminally ill person's family consists of at least two people other than himself or herself, and that person acts altruistically, his or her family members can indeed find that he or she has a duty to die. That his proposal cannot work in a family like that of Sarah and Anita only is not a serious problem for Hardwig. His contention is only that such a duty can arise, and it appears to be possible in families of three or more.

What is more problematic for Hardwig is his notion of the family. I may be called on to make great sacrifices for my family, and they will often be my moral duty such that I would be held morally blameworthy if I failed to carry out my duty, but what sort of family could decide that one of their own had a duty to die? What sort of family could ask this of one of their members? These questions, although admittedly more nebulous, are serious threats to Hardwig's duty to die.

Acknowledgments

I would like to thank Margaret Battin, Troy Booher, Susan Downs, Leslie Francis, Bruce Landesman, Karthik Nadeson, Margaret Plane, and Jacqueline Solon for their thoughtful comments and careful reading of this paper. I would also like to thank my wife, Katie, for her unwavering support in this and all of my projects.

Notes and References

[1]There is an important issue to be raised concerning the relationship the family members have with one another. One might object that families could easily be dysfunctional to a point that they cannot or *should not* deliberate concerning whether or not one of their members has a duty to die. Such an objection is not relevant to this essay since my objections arise only after the family has decided to discuss the issue, and thus they will have a relationship that allows and can facilitate such frank and honest discussion. Such an objection would not affect Hardwig's argument either, of course, since his contention is only that a duty to die as he perceives it can prevail for some families, but not all.

[2]John Hardwig (1997) Is there a duty to die? *Hastings Center Report* 27, no. 2: 37.

[3]Cf. Hardwig, p. 38.

[4]Daniel Callahan (1997) Letter to the Editor of the *Hastings Center Report* 27, no. 6: 4.

[5]I would like to thank Bruce Landesman for this example.

[6]Hardwig, p. 38.

[7]Hardwig, *ibid.*

[8]Hardwig, p. 40.

[9]There might even, in two-person families, be a more serious problem for Hardwig. If Sarah is only considering her loved ones (Anita) and Anita is only considering her loved ones (Sarah), they might not be able to reach any conclusion at all.

[10]Hardwig, p. 38.

[11]Briefly, the individualistic fallacy, coined by Hardwig, consists of decision making as though one's self were the only person affected by one's decision. For a more complete explication, *see* Hardwig, pp. 35,6.

[12]I am indebted to Troy Booher for helping clarify this issue.

Index

E

F

G

H

I

J

K

L